Biomedical Ethics Reviews · 1983

Biomedical Ethics Reviews

Editors

James M. Humber and Robert F. Almeder

Board of Editors

Biomedical
Ethics
Reviews

•

1983

Edited by

JAMES M. HUMBER and ROBERT F. ALMEDER
Georgia State University, Atlanta Georgia

Humana Press   ·   Clifton, New Jersey

Library of Congress Cataloging in Publication Data

Main entry under title:

Biomedical ethics reviews 1983.

   Includes bibliographical references and index.
   1. Medical ethics—Addresses, essays, lectures.
I. Humber, James M.   Almeder, Robert F. [DNLM:
1. Ethics, Medical—Periodicals. W1 B615]
R724.B494 1982   174'.2   83-74
ISBN 0-89603-041-5

©1983 The HUMANA Press Inc.
Crescent Manor
PO Box 2148
Clifton, NJ 07015

Printed in the United States of America.

ISBN: 0-89603-041-5

CONTENTS

EUTHANASIA

SURROGATE GESTATION

THE DISTRIBUTION OF HEALTH CARE

THE INVOLUNTARY COMMITMENT AND TREATMENT OF MENTALLY ILL PERSONS

PATENTING NEW LIFE FORMS

Preface

In the past decade the body of literature in the area of biomedical ethics has expanded at an astounding rate. Indeed, on every major topic, the literature in this area has multiplied, and continues to do so, so rapidly that one can easily fall behind important advances in our thinking about and understanding of the problems of contemporary bioethics. Awareness of this need to keep apace of developments in the area prompted a recent reviewer of our earlier collection *Biomedical Ethics and the Law* (Plenum, 2nd edition, 1979) to suggest that somebody ought to offer the service of providing a biennial review or update of the literature on the various central topics in bioethics.

Thomas Lanigan, of The Humana Press, agreed with this last suggestion and so asked us to edit a series of texts consisting of previously unpublished essays on selected topics, a series that would seek to review and update recent literature on the central topics, while also striving to advance distinctive solutions to the problems on the topics under discussion. Accordingly, this first collection of previously unpublished essays focuses on the selected topics, and the authors commissioned were charged with addressing the basic problems assigned while also bringing the reader either directly or indirectly up to date on the relevant literature. In some cases the review of the literature is quite explicitly done and in other cases the authors seek to solve or resolve specific problems in ways that build on the best of recent literature in the area.

On the whole, we believe the purpose of providing a review of the recent literature, as well as of advancing bioethical discussion, is admirably served by the pieces herein collected. We look forward to the next volume in this series and very much hope the reader will also.

James M. Humber
Robert F. Almeder

CONTRIBUTORS

ROBERT F. ALMEDER · *Department of Philosophy, Georgia State University, Atlanta Georgia*

ROBERT L. ARRINGTON · *Department of Philosophy, Georgia State University, Atlanta, Georgia*

ROBERT BAKER · *Department of Philosophy, Union College, Schenectady, New York*

THEODORE M. BENDITT · *Department of Philosophy, University of Alabama in Birmingham, Birmingham, Alabama*

L. B. CEBIK · *Department of Philosophy, The University of Tennessee, Knoxville, Tennessee*

NICHOLAS FOTION · *Department of Philosophy, Emory University, Atlanta, Georgia*

RICHARD T. HULL · *Departments of Philosophy and Medicine, State University of New York at Buffalo, Amherst, New York*

JAMES MUYSKENS · *Department of Philosophy, Hunter College–CUNY, New York, New York*

LISA H. NEWTON · *Department of Philosophy, Fairfield University, Fairfield, Connecticut*

JAMES RACHELS · *Office of the Dean, School of Humanities, University of Alabama in Birmingham, Birmingham, Alabama*

Topics scheduled for review in future volumes of

BIOMEDICAL ETHICS REVIEWS

Public Policy Formation Concerning Human Research
The Ethics of *In Utero* Surgery
Embryonic Cryogenics
Genetic Screening
Occupational Health
The Right to Health in a Democratic Society

Section I

Euthanasia

Section I

Ecthyma

Introduction

Professor Robert Baker begins his discussion in "On Euthanasia" by briefly reviewing the history of the euthanasia debate. Next, Baker succinctly outlines the contemporary argument for euthanasia:

1. Killing/letting die are morally equivalent concepts.
2. It is contemporary medical practice to allow some patients to die.
3. Therefore, killing those patients who are currently allowed to die does not constitute a morally significant change in medical practice.
4. Furthermore, letting patients die has not generated any of the abuses envisioned by anti-euthanasiasts.
5. Because killing puts an end to pain and suffering it is morally preferable to being let die.
6. Therefore it would be better to kill the patients who presently are being allowed to die.

Baker attacks every premise in this argument. He argues that there are morally significant differences between killing and letting die, and that it is *not* contemporary medical practice to allow some patients to die. Thus, on Baker's view, killing patients would constitute a morally significant change in current medical practice. Furthermore, Baker argues that mistakes and the potential for abuse are far greater than euthanasia advocates realize. In addition, he contends that means presently are available for ensuring a completely pain-free existence for terminally ill patients. Given all of these factors, then, Baker concludes that euthanasia should not be permitted.

In "The Sanctity of Life" Professor James Rachels attempts to develop a new defense of euthanasia—one based upon the principle of the sanctity of life. Rachels begins his discussion by briefly comparing the eastern and western traditions on the doctrine of the sanctity of life. After noting the differences between the two positions, Rachels argues that both traditions must be rejected because each falls victim to the same error. Specifically, neither tradition recognizes that there are two very different senses in which the term 'life' may be used: (a) When we speak of 'life' we may mean to say that something is *alive,* or (b) we

3

may mean that a being *has a life,* i.e., that he or she lives, or is capable of living, in a certain (qualitative) way.

Rachels contends that in formulating their principles of the sanctity of life, both the eastern and western traditions have been more concerned with *being alive* than with *having lives.* And for Rachels, this is a mistake. Thus Rachels argues that the principle of the sanctity of life should be revised so as to protect those who have lives rather than those who are simply alive. If the principle of the sanctity of life were reformulated along these lines, Rachels aknowledges that it would not provide a clear-cut answer as to how we should act in all cases where life and death decisions had to be made. Nevertheless, he is clear that such a reformulated principle would require at least: (i) that the lives of some higher animals (e.g., monkeys) be given more protection than now accorded, and (ii) that active euthanasia be permitted for at least some human beings.

On Euthanasia

Robert Baker

> One dies just as it comes; one dies the
> death that belongs to the disease one has
> (for since one has come to know all dis-
> eases, one knows too that the different le-
> thal terminations belong to the diseases
> and not to the people; and the sick person
> has so to speak nothing to do).
>
> **R. M. Rilke**
> *The Notebooks of Malte Laurids Brigge*

The Traditional Argument

Two insights motivate proponents of euthanasia. One is that pain and
all the other assaults upon human autonomy attendant upon disease can
be so severe that it can be more humane to kill than to prolong dying or
even to cure. The second insight is that our lives are preeminently our
own to do with and terminate as we will. Some may wish to challenge
these insights—they were certainly controversial at other times and
places[1]—but neither the traditional nor the contemporary philosophical debate on euthanasia has focused on them. Although euthanasiasts
sometimes act as if this were the issue, antieuthanasiasts by and large
concede the point and take the issue to turn not on the moral propriety
of isolated acts of mercy killing (on the proverbial desert island), but
on the propriety of accepting euthanasia as part of public morality,
sanctioned by law. The antieuthanasiasts' point is that, just as the mere
immorality of an act does not suffice to justify making it illegal (which
is why there are no statutes proscribing nastiness), so too the mere hu-
maneness and decency of an act does not suffice to provide a warrant
for its legality. For laws regulate not individual acts but social prac-

5

tices; and practices, by nature, are iterative; and extensive iteration necessarily transforms the improbable consequences of individual acts into virtual certainties.

The *locus classicus* for the traditional antieuthanasia argument is a remarkable review of Glanville Williams' book, *The Sanctity of Life and the Criminal Law,* that Yale Kamisar wrote for the *Minnesota Law Review*[2]; the classical euthanasiast reply was written as a rejoinder by Williams a few months later.[3] The gist of Kamisar's objection to legalizing euthanasia is given early on in his analysis:

> Williams' . . . proposal . . . raises too great a risk of abuse and mistake to warrant a change in the existing law. That a proposal entails risk of mistake is hardly a conclusive reason against it. But neither is it irrelevant. Under any euthanasia program the consequences of mistake, of course, are always fatal. As I shall endeavor to show, the incidence of mistake of one kind or another is likely to be appreciable. If this indeed is the case, unless the need for the authorized conduct is compelling enough to override it, I take it that the risk of a mistake *is* a conclusive reason against such authorization. I submit too that the possible radiations from the proposed radiations, e.g., involuntary euthanasia of idiots and imbeciles . . . and the emergence of the legal precedent that there are lives not "worth living," give additional cause to pause.
>
> I see the issue, then, as the need for voluntary euthanasia versus (1) the incidence of mistake and abuse; and (2) the danger that legal machinery designed to kill those who are a nuisance to themselves may some day engulf those who are a nuisance to others.[4]

In his response to Kamisar, Williams reiterates the traditional argument for euthanasia.

> The argument in favor of voluntary euthanasia in the terminal stages of painful diseases is quite a simple one, and is an application of two values that are widely recognized. The value is the prevention of cruelty. Much as men differ in their ethical assessments, all agree that cruelty is an evil . . . Those who plead for the legalization of euthanasia think that it is cruel to allow a human being to linger for months in the last stages of agony, weakness and decay, and to refuse him his demand for merciful release. There is also a second cruelty . . . the agony of relatives in seeing their loved ones in this plight. . . . The second value involved is liberty. . . . And . . . the liberty involved is that of the doctor as well as the patient.[5]

The suggestion that the law against murder limits the liberty of physicians to care for their patients introduces a new theme into the euthanasiast's argument (one that finds a sympathetic resonance among some physicians). On perhaps a more ominous note is the chord that Williams sounds in dismissing the German experience with euthanasia.

The German "eugenic euthanasia" of twelve million "unfit" individuals ("the holocaust") grew out of the eugenic and euthanasia laws of the pre- and early Nazi years. It started as the legalization of voluntary euthanasia—and was originally promoted on the basis of the traditional euthanasiast arguments. But over the course of a few short years the humanely motivated voluntary euthanasia was transformed into a program for the involuntary "euthanasia" of those with severe physical and mental disabilities, and then into a program of ageicide, senilicide, and finally into a program of racial genocide.

Kamisar found the transmogrification of a humane euthanasia program into the holocaust daunting—to say the least. Williams' response—which again represents the main line of euthanasist argument—is that this transformation was a social artifact peculiar to the politics of Nazism. Thus he suggests that each issue should be "debated on its own merits" and questions whether Kamisar seriously fears:

> that anyone in the United States is going to propose the extermination of people of a minority race or religion. Let us put aside such ridiculous fantasies and discuss practical politics.[6]

Curiously, Williams himself then raises some of the very specters he dismisses as "ridiculous." Thus he himself suggests that voluntary euthanasia programs would serve as a wedge opening opportunities for "a body of opinion" that would favor "the involuntary euthanasia of hopelessly deformed infants."[7] Williams also toys with the idea of geriatricide.

> . . .the fact that we may one day have to face is that medical science is more successful in preserving the body than in preserving the mind. It is not impossible that, in the foreseeable, medical men will be able to preserve the mindless body until the age, say, of 1000, while the mind itself will have lasted say, only a tenth of that time. What will mankind do then? It is hardly possible to imagine that we shall establish huge hospital-mausolea where the ages are kept in a kind of living death. Even if it is desired to do this, the cost of the undertaking may make it impossible.

This is not an immediate problem, and we need not face it yet. The problem of maintaining persons faced with senile demen-tia is well within our economic resources as the matter stands at present. . . . Perhaps, as time goes by . . . men will become more resigned to human control over the mode of termination of life. Or the solution may be that after the individual has reached a certain age, or a certain degree of decay, medical science will hold its hand and allow him to be carried off by natural causes. But what if these natural causes are themselves painful? Would it not be better kindness to substitute human agency?[7]

In retrospect, Williams' vision of medical progress as evoking the nemesis of hospital mausolea seems prescient. He sensed the problem well before the development of the modern intensive care unit; he wrote almost two decades before that fateful day in 1975 when Karen Ann Quinlan entered bioethical history by demonstrating that the med-ical complications consequent to using the MA-1 respirator can be less significant than the moral and legal complexities. Not surprisingly, the themes he sounded here—about the expense and indignity of the end-less prolongation of physical existence by means of medical technology—came to play an ever larger part in more recent statements of the traditional argument.

The Contemporary Philosophical Debate

In the past two decades the traditional argument for euthanasia has been developed with varying degrees of eloquence and philosophical acumen by Anthony Flew,[8] Marvin Kohl,[9] and O. Ruth Russell,[10] among others. In recent years, however, the contemporary philosoph-ical debate has taken a rather different twist. James Rachels, Peter Singer, Michael Tooley, and other philosophers not only urge the tra-ditional arguments for euthanasia, they also maintain that the tradi-tional response to this argument (epitomized by Kamisar) has been re-futed by contemporary medical practice.[11] For, they argue, the technology of intensive care has forced physicians to adopt a policy of letting patients die—a policy that is morally indistinguishable from mercy killing, but that generates *none* of the abuses envisioned by Kamisar and other antieuthanasiasts. Thus, while the classical prob-lems of the pain and indignity of disease and lingering death remain, the fear that mercy killing is a dangerous solution to these problems has been demonstrated to be unfounded. Therefore, these philosophers suggest, only intellectual Luddites or moral atavists will maintain their arcane objections to mercy killing in the face of the evident fact of human suffering.

The contemporary case against those who object to euthanasia takes the following form.

1. Killing/letting die is a morally irrelevant distinction. (Or, as the point is alternatively put, the concepts are morally equivalent.)
2. It is contemporary medical practice to allow some patients to die.

3. Therefore, killing those patients who are currently allowed to die does not constitute a morally significant change in medical practice.
4. Furthermore letting patients die has not generated any of the abuses envisioned by antieuthanasiasts.
5. Killing is morally preferable to being let die.

6. Therefore it would be better to kill the patients who are presently being allowed to die.

Philosophical attention, naturally enough, has been largely riveted on premise (1), i.e., upon the purported moral equivalence/irrelevance of the killing and letting die distinction. Premise (1) originates from the argument that there is no *intrinsic* or *inherent* difference between an act or an omission, between killing and letting die. It was directed against some naive antieuthanasist who held the view that the bare difference between killing and letting die—the conceptual difference, as it were—sufficed to show that killing is worse than letting die.

To disprove this position Bennett, Rachels, and Tooley developed the argument from comparable cases. This argument rests on comparison of the case of K, who has intentionally killed V for evil reasons, with that of L who has intentionally let V' die, for the same evil reasons. The victims V and V' are both dead; the evil intent is the same; how then can L be any less evil than K? Surely the acts are equally reprehensible. "By their fruits ye shall know them"; since the fruits of both acts are equally bitter, how can they be anything but moral equivalents? How then can killing, considered in and of itself, be morally worse than letting die? Obviously, therefore, the killing/letting die distinction is irrelevant to gaging the morality of acts.

It is not at all clear that anyone ever held the position this argument was designed to refute. James Rachels takes the AMA to hold this position because they endorse the cessation of extraordinary treatment, while condemning mercy killing; but, as we shall see, the AMA was not claiming that the bare difference between killing and letting die, or between an act and an omission, constitutes a significant moral distinction. Some philosophers, perhaps most notably Phillipa Foot,[12] have argued that negative duties, such as the duty not to kill, are comparatively stronger than, and hence override, positive duties, such as the duty to relieve suffering. Since the argument from comparable cases does not directly address the question of duties, it is not immediately evident that it counts as a counterexample to Foot.

The virtues of the argument from comparative cases as a counter to the claim that killing is intrinsically, inherently, or, in and of itself, worse than letting die, is of less interest to us here, however, than subtle transformation of this argument into a positive case in favor of euthanasia. If, after all, killing is not intrinsically different from letting die; if there is no moral difference between the acts themselves; if, as acts, they are morally equivalent; then if it is morally proper to let die, it is also morally proper to kill. Thus the argument from comparative cases appears to generate not only a critique of an antieuthanasiast argument, but a positive case for euthanasia—that is a case for euthanasia in any circumstance in which it is morally appropriate to allow someone to die.

Beauchamp[13] and Steinbock[14] have made the point that the argument from comparable cases merely establishes the equivalence of results in an individual case and will not establish the moral equivalence of concepts *per se*. Suppose, for example, that a villain K kisses V with the evil intent of thereby killing V; suppose further that this is, indeed, the kiss of death; so that V dies. Does it follow from the fact that *in this one case* the consequence of kissing is death that kissing is—in and of itself, intrinsically, or in general—the moral equivalent of killing? Does this argument establish that kissing the bride is intrinsically, or, in general, the moral equivalent of killing the bride? Are all kisses the moral equivalent of the kiss of death?

Although I should not like to rehearse the entire literature on the subject, it must be remarked that Beauchamp and Steinbock have not had the final word. Both Rachels and Tooley[15] have attempted to develop a criterion of moral equivalence that will show that where consequences and intent (did the bride intend to kill the groom by her kiss?) and reprehensibility are the same, the acts are morally equivalent. The debate is thus very much alive.

The reason why further rehearsal of the controversy is not appropriate in the present context is that there are two striking differences between killing a patient and allowing a patient to die: killing a patient can be swift and painless, but it is also irreversible; letting a patient die can be prolonged and painful, but the decision can be reversed and can, in an odd sort of way, be self-falsifying. To put these two points into focus, reflect on the fact that to allow, or to let, is to forbear preventing. Thus allowing and letting presuppose the possibility of preventing something, and suppose that this possibility is not actualized. One cannot, for example, be said to have "let a cat out" if the cat is out already, or if the cat's exit—like a sunset—is beyond one's powers to prevent. So, to allow someone to die is to be able to prevent the person's death, but to opt not to do so. It follows that lettings die are reversible as long as the (physician's) option for medical intervention can still be exercised. Sometimes, if physicians misprognose, a patient may be allowed to die, but persist in living. (Actually, it is improper to call these cases allowances of death since there was no death to allow—i.e., one of the presuppositions of allowing has not been satisfied.) Were these patients killed, of course, they would be dead—whether this is to the good, or not, will depend upon whether they survive to a life worth living.

These are, as it turns out, morally significant differences between the acts of killing and allowing to die. Any theoretical analysis that can not accommodate these differences is wrong; and if the differences are accommodated then, for all practical purposes, killing and letting die must be regarded as morally non-equivalent with, as it were, different advantages and disadvantages.

The Virtues of Killing

The problem with allowances of death is that they can be prolonged and possibly painful. And it is on these grounds that euthanasiasts urge the moral superiority of killing [premise (5)]. In the context of medical practice the euthanasists do *not* take killing to be the moral equivalent of letting die; rather they hold killing to be the morally preferable to allowing death because it can be swift and painless. Peter Singer puts this point quite well in his beguilingly accessible book, *Practical Ethics*, when he criticizes the practices of the British physician John Lorber:

In an article in the *British Medical Journal*, John Lorber has charted the fate of 25 infants born with spina bifida on whom it has been decided, in view of the poor prospects for a worthwhile life not to operate. . . . Lorber freely grants that the object of not treating infants is that they should die soon and painlessly. Yet of the 25 untreated infants, 14 were still alive after one month, and seven after three months. In Lorber's sample all the infants die within nine months, but this can not be guaranteed. An Australian clinic following Lorber's approach to spina bifida found that of 79 untreated infants, 5 survived for more than two years. . . .

If we are able to admit that our objective is a swift and painless death we should not leave it to chance to determine that this objective is achieved. Having choosen death we should ensure that it comes in the best possible way.[16]

The Moral Propriety of Letting Patients Die

Perhaps, however, one should not choose to let these patients die. Although it is a bit of a digression from the immediate question of the comparative merits of killing as opposed to letting die, it is worth entertaining the question of whether it is reasonable to let these babies die. Perhaps the conclusion that one should draw from this case is that neither killing nor letting die are morally acceptable practices. Surely Lorber's practices are suspect. Recall Kamisar's warning that the "machinery designed to kill those who are a nuisance to themselves may some day engulf those who are a nuisance to others."[17] Is it not possible that Lorber is allowing these babies to die because they are a nuisance to their parents and an expense to the British National Health Service (and hence to the British taxpayer)?

Dr. John Freeman, of Johns Hopkins Medical School, has argued that had these children received the same treatment at Lorber's clinic in Sheffield that they would receive in Baltimore, it is reasonable to expect that only 17% would die, that 35% of the survivors would have a below normal IQ, and that 28% would be unable to walk.[18] Given these percentages, is it clear that these children are better off dead? Is it so evident that it is better to be dead than retarded? Most retarded people can talk. (All but a handful of chimps are mute.) Are these retarded people better off dead simply because they can never learn to read? (If so, would almost all chimps be better off dead?). Is it really obvious that a low IQ makes life worthless? What of dogs and cats whose IQ is less than that of almost all retarded humans? Perhaps it is the fact of being confined to a wheel chair that is significant. Suppose, however, that you were faced with the choice: death versus a chance of retardation (odds, 1:3 in favor of normality) and/or severe disablement (odds, 1:5 in favor of normality). Given the overwhelming odds in favor of normality, would you still choose death? Suppose that normality were marred by the burden of a "shunt" (a plastic tube that circulates fluids that is frequently implanted in the heads of people born with myelomeningocele—the primary cause of spina bifida). The shunt requires a certain degree of medical attention; it potentiates headaches, and, in a minority of patients, it requires frequent corrective operations. So the "normality" offered a person with spina bifida is, frequently, a bit more fragile and less free than that of the healthy. If one were facing the choice of death as opposed to better than even odds for this fragile life, would you consider yourself better off dead?

Even if *you* would choose certain death over a possibly abnormal life there remains the question whether, as Dr. Freeman puts it,

> . . .we have any right or ability to assess the quality of life of another?
>
> Perhaps with an adult we could project ourselves into that situation and guess how we would feel if that would be our quality of life. But there are many qualities of life that I would not choose, qualities of life with or without physical disability. I might not choose to grow up retarded, or in an underdeveloped country, or underpriviledged in our own country. But there are many who grow up in these situations who would rather have those handicaps than be dead.
>
> The child growing up with a disability or handicap has a far different perception of his quality of life than an adult who suddenly becomes disabled or handicapped. Do we even have the ability to project ourselves into the child's situation?[19]

Lorber, of course, suggests that he can project himself into the child's place and determine with reasonable accuracy that the child would chose not to live, and hence is best off dead. One need not be a committed antieuthanasiast to suspect, however, that Lorber's decision is influenced by economics of the National Health Service and the belief that *parents* are better off if these children are dead.

The Generalized Argument From Medical Practice

The selective nontreatment of birth defective children advocated by Lorber is controversial in Britain, and neither accepted nor practiced in the US. It therefore provides no grounds upon which one might argue for the legalization of euthanasia. Certainly, one would not seriously propose the legalization of euthanasia in order to regularize a suspect treatment of children with a rare birth defect. If the medical allowance of death is to provide a basis for a euthanasiast argument, then it must be shown that there is a large body of patients that physicians take to be better off dead—and allow to die. James Rachels believes that there is just such a class of patients and argues for mercy killing on this basis.

> Physicians often allow patients to die. . . . For example, a doctor may leave instructions that if a hopeless, comatose patient suffers cardiac arrest, nothing is to be done to start his heart beating again. "No-Coding" is the name given to this practice, and the consent of the patient and/or his immediate family is not commonly sought. This is thought to be a medical decision (in reality, of course, it is a moral one) which is the doctor's affair.[20]

One point is undeniable: were "no-coding" a form of allowing patients to die, it would indeed be standard American medical practice for physicians to let large numbers of patients die; for no-coding is, indeed, a routine practice in intensive care units (ICUs) and other wards of American hospitals.

But is Rachels correct in taking no-coding to be a case of allowing patients to die? A "code," it must be remarked, is hospital language for an emergency. "Code One Pharmacy" might be an alert that there is a fire in the pharmacy. To call a "code" on a patient is to announce that there is a medical emergency—typically a cardiac or respiratory arrest—that must be dealt with immediately. To say that a patient is "no-coded" is to say that this emergency treatment will not be given. Hence, it is *prima facie* plausible to claim, as Rachels does, that no-codes constitute allowances of death.

To make the argument for the moral preferability of mercy killing work, however, Rachels must argue that no-codes are allowances of death initiated in Lorber-like fashion, on the grounds that these patients are better off dead. The published protocols that stipulate unit and hospital policy for no-codes (and other "cessations" of treatment), however, suggest that the perceived justification of no-code is quite different. Thus all published protocols state that the necessary condition for a no-code is a prognosis of terminality unresponsive to medical treatment. There are two prototypical no-coding protocols (in the sense that all other protocols appear to be variants of these two). One is published by the intensive care units at Beth Israel Hospital (Boston), the other by Mount Sinai Hospital (New York). The former specifies that no-coding is appropriate only

> . . .if a medical judgment is reached that a patient is faced with such an illness and imminence of death that resuscitation is medically inappropriate.[21]

The statement continues on at some length, clarifying, among other matters, what is meant by 'irreversibility'

no known therapeutic measures can be effective in reversing the course of illness.

The Mount Sinai protocol, and others that have been modeled upon it,[22] echo the theme of medical futility as a prerequisite for terminating treatment. Cessation of treatment is appropriate for patients

. . .whose clinical course has deteriorated to the point that there is every expectation that he will not survive despite maximal therapeutic efforts. . .

(defined as a chance of survival of less than one in 100,000).
In the case of these patients

No active or specific therapy (transfusions, antibiotics, etc.) *to be initiated.* Conservative, passive medical care replaces heroic measures. Relief of suffering is primary goal. Mechanical therapeutic measures (volume ventilators, cardiac pacemakers) already in use are continued, but not initiated *de novo*. No emergency resuscitation will be initiated. Particular attention is given to comfort, including oxygen therapy for shortness of breath, intravenous fluids to prevent excessive thirst, and analgesics for relief of pain.

A 1973 policy statement by the AMA resonates the necessity of a prognosis of untreatability as the *sine qua non* for nontreatment endorsing

. . .the cessation of the employment of extraordinary means to prolong the life of the body when there is irrefutable evidence that biological death is imminent. . . .[23]

Similarly the American Heart Association Guidelines for terminating cardiopulmonary resuscitation state that termination is appropriate when the patient fails to respond to treatment and so is deemed "unresuscitatable."[24] Although other notes are occasionally sounded, the dominant theme in all published policy statements on cessation of treatment, including no-coding, is that treatment should cease only when it is medically futile. Rachels appears not to appreciate this point—remarking on the 1973 AMA statement "what is the cessation of treatment, in these circumstances, if it is not the intentional termination of the life of one human being by another?"

Of course it is exactly that, and if it were not, there would be no point to it.[25]

Is the intentional termination of human life the only possible point to no-coding? Imagine that someone is drowning at sea. You throw them a life jacket. It falls short. You throw them another, and another. But they drift further and further away. Finally the futility of further attempts to reach them with a life preserver is so evident that you cease to throw any. Is it your intent that this person die? Rachels would appear to think so. He seems to believe that "cessation" of rescue when there is "irrefutable" evidence that drowning is imminent despite any possible attempt at rescue, must be the intentional termination of life, "otherwise there would be no point to it." But the pointlessness is, as it were, on the other foot. It is pointless (and expensive) to continue an attempt at medical rescue when, to paraphrase the Beth Israel protocol, "no known rescue measures can be effective in reversing the course of drowning"; or when to use the words of the Mount Sinai protocol, the person's "course has deteriorated to the point that there is every expectation that he will not survive despite maximal rescue efforts."

Suppose that people were frequently swept overboard and that when they were spotted a "code"—or rescue effort—was called. However, if a rescue attempt was believed to be futile, a "no-code" was called. "No active or specific rescue attempt was engaged in (life preserver throwing was not to be initiated). Conservative, passive care replaces heroic measures. Relief of suffering is the primary goal (to paraphrase the Mount Sinai protocol)—perhaps vials of morphine were floated out to people, or some such thing; certainly a careful watch was kept on them. But "no emergency measures were initiated." Would such a no-code constitute an intentional termination of life? Of course not. But this is just what a no-code is—the cessation of a futile attempt at medical rescue. And to cease saving in this way is not tantamount to intentionally terminating life.

Is it morally preferable for people who are assigned no-codes to be killed? That after all is the euthansiast's point—why let someone die/drown slowly and, presumably, painfully if they can be killed quickly and painlessly? But is the drowning person better off dead? Are they not best off rescued? No-codes are reversible and—in point of fact—are frequently reversed. In one of the few published studies of the effects of no-coding, Dr. Ake Grenvik and his associates at Pittsburgh Presbyterian Hospital reported that 8.6% of all no-coded patients later left the hospital alive.[26] (If 8.6% of all those swept overboard and "no-coded", but later drifted back into rescue range, would any no-coded person wish to be shot?) In a neonatal intensive care unit studied by the author, three no-codes were written during a three-month period, and all three happened to survive in good health. As it happened, none of these infants suffered cardiac or respiratory crises during the period of time they were no-coded and all recovered from their cerebral insults (two cases) or outgrew their extreme prematurity, and so became potential codes. (All three are now alive and healthy.)

The fallibility of the prognoses underlying no-codes thus contrives to make their status as uncertain achievers of death a virtue rather than a vice. Even if the purpose of no-coding were the intentional termination of life, as Rachels believes it to be, if the prognosis that recovery is less than one chance in 100,000 (the standard precondition for a no-code) is accurate only 91.4% of the time (to use the Pittsburgh Presbyterian figures), then the level of *inaccuracy* for these prognoses is 8.6%. And, an 8.6% level of prognostic inaccuracy would render killing's guarantee of death a questionable virtue. Recall that the point of no-coding is not to terminate life, but to cease futile rescue. If there is an 8.4% chance of survival, the achievement status of killing is a vice. It is *prima facie* morally preferable to be no-code rather than to be killed—however mercifully the killing is intended—because it is *prima facie* better to be alive than dead.

One final note on Rachels' claim that no-coding is a form of letting die. A no-code is called only when lifesaving medical interventions are deemed futile—i.e., when it is believed that there is no possibility that medical intervention can save the patient's life. Allowings and lettings, as we have argued, presuppose the possibility of preventing. Since no-codes are called only when death prevention is presumed to be impossible, no-codes cannot be thought of as allowances of death. Indeed, when on *post mortem* autopsy a medical unit discovers that a patient's death was indeed preventable, the no-code will be taken to have been improper. No-coding is proper, therefore, only when it is *not* an allowance of death. In so far as no-coding is an allowance of death, it is improper. So even if one could somehow show that killing was the moral equivalent of allowing to die, the prevalence of the practice of no-coding would still provide no case for mercy killing. For no-coding is not allowing to die.

The Traditional Argument Reconsidered

The contemporary attempt to circumvent the traditional arguments surrounding the legalization of euthanasia rests on two premises: (1) the supposition of the moral equivalence/irrelevance of the killing/letting die distinction, and (2) the premise that it is contemporary medical practice to let patients die. Neither of these premises have proved correct. Killing and letting die are not morally equivalent, nor is the distinction irrelevant; if only because killing guarantees death, where letting die is reversible. Since it is simply not true that it is contemporary medical practice to allow patients to die, the possible preferability of killing to allowances of death provides no basis for the legalization of medical mercy killing. In sum, the line of argument urged by Singer, Rachels, Tooley, et. al.—the attempt to obviate the traditional debate by a bit of conceptual slight-of-hand that distracts our intellectual attention by focusing it on acts and omissions, killings and lettings die, rather than the risks and rewards of organized mercy killing—must be recognized as bankrupt. If there are good reasons to legalize mercy killing, they will, in the end, be the traditional reasons, and if the traditional arguments for euthanasia are not compelling, then the legalization of mercy killing cannot be justified.

Mistake and Abuse

Mistake and abuse are, as Kamisar points out, the two primary reasons for fearing the legalization of voluntary euthanasia. Abuse, or rather the potential for abuse—the possible ways in which a voluntary program could be corrupted—has been the dominant theme of the traditional debate. One overriding concern of the antieuthanasiasts has been that voluntary will ultimately be transubstantiated into an involuntary program. We shall not say much about the slippery slope as a potential abuse of an institutionalized program of legal voluntary euthanasia—except to rehearse again Kamisar's observation that one would have fewer fears of such a transmogrification were the partisans of voluntary euthanasis not equally committed advocates of involuntary euthanasia. One would be a bit more reassured when Singer and Williams dismiss fears about the ''slippery slope'' from voluntary to involuntary euthanasia if Singer were not also a proponent of killing Down's babies, children with myelomeningoceles, and hemophiliacs, and if Williams were not toying with the idea of involuntary geriatricide. Speculating about the slippery slope, however, is a species of futurology that yields results so suspect that only the previously converted can accept them with any degree of equanimity. There are more straightforward reasons for worrying about the potential for abuse, which will become apparent when we consider mistakes.

Mistakes—the possibilities of mistakes in prognosis and diagnosis—provide a more readily quantified mode of assessing the dangers of legalizing voluntary euthanasia. Suppose, for purposes of this analysis, that it is reasonably accurate to assume that industralized North America has a population of 300,000,000; that the average person has contact with a medical practitioner 3-⅓ times a year (one billion visits a year), that some 50,000,000 people are in-patients at a health care institutions (including nursing homes) annually, that about 3,000,000 people die annually, and that 2/3 of these are under some form of medical care when they die.[27] Assume, finally, that half of the 2,000,000 who die in the care of physicians will experience a significant degree of apparently unremediable pain. Given all of these assumptions, the maximal population that could benefit from the legal institutionalization of a voluntary euthanasia program would be 1,000,000 people annually.

Presumably a conservative voluntary euthanasia program will restrict the option of being mercy killed to hospitalized patients with a prognosis of irremediable pain and irreversible terminality (i.e., in fewer than one case in 100,000 would treatment prevent death). Assume that the prognosis of irreversible terminality is 99% accurate and that one-third of all patients so prognosed opt to be mercy killed. Under these circumstances, ten thousand people (1% of the one million prognosed terminal) will be misprognosed and so the third of these who opt for euthanasia (3,333) will be killed unnecessarily—in order that 330,000 others can escape a presumably painful death. *Since the benefits conferred are—in some crude sense—one-hundredfold greater than the price in unwarranted deaths, the possibility of misdiagnosis at this level of prognostic accuracy does not appear to provide any significant objection to institutionalizing legal euthanasia.* Moreover, a person faced with a 99 chances in 100 of a painful death, as opposed to a certainty of a painless one, would be well within the bounds of reason in opting for the painless alternative. Thus voluntary euthanasia appears to be a reasonable choice in situations of almost absolute prognostic certainty.

In reality, however, the likelihood of misprognosis is probably on the order of 10% (recall the 8.4% inaccuracy of no-code prognoses). If prognoses are only 90% accurate under a voluntary euthanasia program, for every million patients prognosed irreversibly terminal, 100,000 will experience a remission of illness. If one-third of these opt for euthanasia, there will be 33,333 unwarranted mercy killings. Roughly one unwarranted killing for every ten that are justifiable. Given this level of prognostic inaccuracy, the rationality of opting for euthanasia is unclear. For although the physicians prognose certain death, since their prognoses are only 90% accurate, one still has a one in ten chance of living. Hence, save where it is a question of excrucia-

ting pain, it is doubtful that it is better to opt for certain death in preference to a one in ten chance for life. Leaving aside the question of pain,
then, it should be clear that more fallible the medical prognosis, the
less rational it is institutionalize voluntary euthanasia.

Although rational euthanasia is compatible with a certain degree
of misprognosis, any degree of abuse impeaches the enterprise. In a
conservative system of voluntary euthanasia, mercy killing would be
restricted to hospitalized patients. Thus only the 50 million patients in
hospitals and nursing homes would be at risk to abuse. If such a system
were 99% abuse free, the 1% killed without proper warrant would
number 500,000—or a third *again* as many as die appropriate mercy
deaths. Even a system that was 99.5% abuse free, and which had 90%
accuracy in prognoses, would generate about as many unwarranted
"mercy deaths" as justified killings. The populations at risk are so
large, and the consequences of abuse so significant, that *any* degree of
abuse at all will be sufficient to convince anyone that it is imprudent to
institutionalize legal euthanasia. Which, no doubt, is why opponents
of euthanasia tend to focus so closely on the question of abuse; and
why no large industralized society—other than Nazi Germany—has
experimented with legal euthanasia.

The Question Of Pain

Pain is the driving force that powers the euthanasia debate. Death with
dignity in preference to prolonged depersonalization through pain is
the goal. Peter Singer, for example, cites the case of Derek Humphry,
who abetted the death of his wife Jean.

> This is not a case from the period before effective pain-
> killers . . . Humphry died in 1975. . . . if there were anything
> else that could have been done for her, her husband, a well-
> connected Fleet Street journalist, would have been better placed
> than most to obtain it. Yet Derek Humphry writes:
>   . . . when the request for help in dying meant relief from
>   relentless suffering and pain and I had seen the extent of this
>   agony, the option simply could not be denied . . . And cer-
>   tainly Jean deserved the dignity of selecting her own ending.
>   Perhaps one day it will be possible to treat all terminally ill
> patients in such a way that no one requests euthanasia and the sub-
> ject will become a non-issue; but this still distant prospect is no
> reason to deny euthanasia to those who die in less comfortable
> conditions.[28]

For Singer then, absent the problem of pain, absent the case for euthanasia.

Is it true, however, that the pain of terminal illness is irremediable? Two physicians, Drs. Robert Sade and Anne Redfern once wrote a response to Rachels in the *New England Journal of Medicine*.[29] They claimed, among other things, that "the physician always has the means to relieve suffering." Rachels was dumbfounded by the claim. He replied that he was "astonished" to "read that a terminal illness 'need never be painful' and that 'a physician always has the means to relieve suffering'." To prove this point he cited the case of a young man, Jack, movingly described by the *journalist,* Stuart Alsop, who

> . . . was given an injection to relieve the pain every four hours, but the effects would begin to wear off in half the time. Then he would begin to moan, or whimper very low . . . to howl like a dog. . . . Since such situations do exist I am simply astonished to read . . . that a terminal illness 'need never be painful'. . . .[30]

The case of the unfortunate Jack does not, of course, disprove the physician's claims for the efficacy of modern palliative care. They had claimed that terminal patients *need* never feel pain, *not* that patients did not undergo (unnecessary) suffering. Patients may suffer unnecessarily, just as they are sometimes given unnecessary operations and are sometimes treated incompetently.

The hidden premise in Rachels' reply is his assumption that *if the physicians in question could have prevented this pain, they would have done so.* The reader is thus invited to draw the conclusion that since the physicians did not prevent this pain, they could not do so. Laypeople, including euthanasiasts, can not imagine that physicians (in this case physicians at the National Institute of Health) would allow their patients to suffer from pain—pain that would make a patient moan and whimper and howl like a dog—if the means existed to prevent the pain. Nonetheless, as Alsop discovered by talking to younger doctors, there are indeed effective palliatives. Thus he writes in the penultimate paragraph of his article

> I see no reason on earth why a young man like Jack should suffer one moment of the agony which cheap drugs *now available* . . . can prevent.[31]

Rachels does not take note of this point, perhaps because he refuses to credit it. Yet the "cheap drugs" referred to by Alsop exist, the pharmacopia is standard (morphine, cocaine, etc.), but the procedures by which they are administered are new. The innovations in palliative pharmacology were developed as part of the Hospice movement. (A movement dedicated to the care of terminal patients, that has developed an ideology of terminal care that emphasizes palliation.) The standard way of administering analgesics is *pro re nata,* PRN—on (the patient's demand). The anestheseologists working with the hospice movement, however, find a significant distinction between the methods appropriate to the treatment of acute and chronic pain. Long-term patients suffer from chronic pain; they thus come to recognize and anticipate the course of the pain experience. Such psychological states, the hospice anesthesiologists argue, potentiate pain receptivity, and so exacerbate pain. If painkillers are administered PRN, then, since there is a lag between the time when pain is experienced and the time the drugs become effective, the psychological exacerbation of pain rceptivity cued by the experience of pain will almost always create a greater level of pain, and minimize the efficacy of the analgesic. Effective palliation therefore must start with very high doses that are slowly diminished (''titrated down'') until they enter ''gateway'' between the level that causes sedation and strong enough to achieve effective analgesia.[32]

The best evidence available confirms that the extraordinary effectiveness of these techniques. In 1970—five years *before* the euthanasia of Jean Humphry, more than a decade before Rachels urged the case for euthanasia in the *New England Journal of Medicine*—the Church of England's Board for Social Responsibility set up a Working Party to study the palliative effectiveness of hospice care. The result was a well-documented study of the 577 terminal cancer patients who had been admitted to St. Christopher's Hospice over the course of a given year. Some 367 of these patients experienced pain either on or after their admission to St. Christopher's. After palliative care was initiated, *all but seven (98%) had a completely pain-free existence.* And 99.8% of all the patients at St. Christopher's were free of pain.[33] After examining the medical evidence, the Working Party found there was no need for legislation legalizing medical euthanasia.

In some ways, the Working Party's most interesting observation (and one which the author can confirm from his own observations at US and British hospices) was that

> . . . analgesic drugs can be administered for months without detriment to the patient or his personality or even activities. This is not . . . a living death . . . but a way of enabling someone to live up to the moment of his death.[34]

Effective palliative care, care that extends meaningful life, is possible for all but a fraction of a percent of terminal patients. Why then is such treatment not being given? Had a new "wonder drug" that banished all pain been invented almost two decades ago, there is little doubt that it would now be universally employed. Why have physicians failed to adopt this new technique of applying old drugs? Two theories have been offered. In an editorial in the *New England Journal of Medicine*,[35] Dr. Marcia Angell, deputy editor of *The Journal* points out that the situation is nothing short of scandalous.

> Few things a doctor does are more important than relieving pain. Yet the treatment of severe pain in hospital patients is regularly and systematically inadequate. One study showed that 73 percent of the patients undergoing treatment for pain continued to experience moderate to severe discomfort. This is not for want of tools. It is generally agreed that most pain, no matter how severe, can be effectively relieved by narcotic analgesics.

Angell cites the work of Twycross, the anesthesiologist for St. Christopher's Hospice, and the findings of nonhospice physicians, and condemns the "parsimonious" use of narcotics administered "pro re nata." She suggests that the reason why physicians are unwilling to administer analgesics effectively is that they are too concerned with the possibility of adverse side effects, particularly addiction.

> What are the facts? Addiction for patients who receive narcotics for pain is exceedingly unlikely; the incidence is no more than 0.1 percent. Even those who develop tolerance and physical dependence are unlikely to become addicted, and withdrawal can be accomplished easily if the painful stimulus is no longer present. The purpose of drugs for these patients is, after all, the relief of pain; "street" addicts, in contrast, take drugs for quite different purposes. . . . It is instructive to compare the low incidence of important side effects with the very high incidence of inadequate pain relief. I can't think of any other area in medicine in which such an extravagant concern for side effects so drastically limits treatment.

Angell closes her editorial with an eloquent plea for change.

> Pain is soul destroying. No patient should have to endure pain unnecessarily. The quality of mercy is essential to the practice of medicine; here, of all places, it should not be strained.

Some 15 years ago the perceptive medical sociologist, David Sudnow, suggested a different, and, in my view, more plausable account of the high incidence of inadequate pain relief. Observing that the 'need' for euthanasia seems to be a direct consequence of the patterns of care and regard for pain which modern medical practice has institutionalized.[36]

Sudnow suggested that

From the physician's standpoint . . . once palliative care is instituted, diagnostic enthusiasm becomes less sustainable. The care of such patients is considered essentially a matter for nursing personnel, and physicians lose their interest in the patient. When the point is reached that the likelihood of an improvement of condition is considered negligible, the activities of diagnosis and consequent treatment lose . . . one of their key functions, namely their ability to allow him to demonstrate his technical competencies.[37]

Physicians thus tend to treat patients who are beyond medical help, in the sense of being incurable, as if they were beyond any help at all, *including palliation*. Pain comes to be seen as a nursing problem.

Significantly, the founder of the contemporary hospice movement, Dr. Cicely Saunders, started her career as a *nurse*. Perhaps had she been educated as a physician and learned to think not in terms of caring for patients, but in terms of treating and diagnosing *diseases*, she too would think in terms of malignancies and stages, of diseases and syndromes, of an interesting case of Hodgkin's, instead of worrying about the patient who was in the terminal stage of an "interesting" varient of Hodgkin's disease. As the poet Rilke has observed, from the viewpoint of institutional medicine, "one dies the death that belongs to the disease . . . the different lethal terminations belong to the diseases and not to the people . . . the sick person has so to speak nothing to do." And so doctors do nothing for the people, if they can do nothing for the disease. Fortunately, Saunders, trained as nurse, tried to find a better way to care for the people who happened to be diseased. She headed a team of experts who perfected the palliative efficacy of known analgesics—and offered the resultant pharmacology to the medical profession. Unfortunately, the profession has, until now, been blind to the persons who bear the diseases, and so has not seen reason to introduce the new pharmacology into medical practice.

Euthanasia Reconsidered

The medical profession's blindness to the personal plight of persons who suffer diseases has led Rachels, Singer, and others to believe in the relative ineffectiveness of modern palliative pharmacology, and hence to propose voluntary euthanasia as the analgesic of last resort. Since the analgesic ineffectiveness of contemporary medicine would appear to be an ideological rather than a technological artifact, what becomes of their case for euthanasia? Legalizing institutional mercy killing is a dangerous experiment, whose justification is suspect even if there were a significant incidence of uncontrollable pain. Pain, however, is controllable, so we are really faced with two questions: If the medical profession persists in ignoring effective analgesia, should we legally empower them to kill terminal patients who request death? If the profession introduces effective analgesia, should we legalize voluntary euthanasia?

The answer to the second question depends in some measure on whether hospitals can achieve the same measure of pain relief (99.8% pain free) as hospices. Certainly if the general introduction of modern palliative care pharmacology approximates anything like this level of effectiveness, given the problems of misprognosis and abuse, no sane society would contemplate experimenting with voluntary euthanasia.

If, on the other hand, physicians are not sensitive enough to their patients to provide effective palliation, then, again, no sensible society would entrust them with the power to kill their patients. All of us, including physicians, are sometimes mesmerized by the mystique of the doctor's calling. Consider, however, another group of medical personnel, orderlies. In his study of the way dying patients were treated at a large public hospital, County Hospital, David Sudnow observed that "patients were left throughout the night in the supply room" in order to save the attendants the "onerous" task of making up a bed[38]; he also remarks that another labor-saving practice at the same hospital was to prewrap living patients in their death shroud.[39] These orderlies were not evil, or even mean. They had their jobs to do. They had to make beds and wrap dead patients in their death sheets for delivery to the morgue. They were so engaged in their tasks, however, that they could not see that they were dealing with people. Bureaucratic dehumanization is a cliche, but it is also a reality. These orderlies could not see people *in extremis,* they could only see tasks.

Physicians, too, seem to have lost the ability to see human suffering; instead, they see diseases, conditions, and syndromes that are to be diagnosed, prognosed, and treated. Perhaps they will respond to the voices of Drs. Angell and Saunders and effectively palliate the pain of their patients. Or perhaps terminal patients will all be removed from hospitals and placed in hospices (where they can receive appropriate palliative care). In either case, there would no longer be any compelling reason to legalize euthanasia. But if physicians persist in their refusal to palliate their patients, if they let them whimper and

howl like dogs because they do not see palliation as an interesting medical problem, then society can no more entrust them with the power to
kill their patients than it would entrust the orderlies of County Hospital
with such a power. No sensible society would entrust the power to
mercy kill to a profession that was no longer sensitive to the meaning
of "mercy."

Acknowledgments

I should like to express my gratitude to my colleagues Professors
Kaminsky and Thomas; the former for his perceptiveness, the latter for
his patience. Both have been invaluable to me.

Notes and References

[1]S.E. Sprott, *The English Debate On Suicide* (Open Court, 1961).

[2]*Minnesota Law Review* 42 (1958) No. 6, May 1958; reprinted in T.
Beauchamp and L. Walters, eds., *Contemporary Issues in Bioethics*
Dickenson, 1978), pp. 308–317.

[3]*Minnesota Law Review* Vol 43, No. 1 (1948); reprinted in Beauchamp and
Walters, pp. 318–328 (ref. 2).

[4]Beauchamp and Walters, p. 309.

[5]Beauchamp and Walters, p. 318.

[6]Beauchamp and Walters, p. 322.

[7]Beauchamp and Walters, p. 322.

[8]Anthony Flew "The Principle of Euthanasia," in A. B. Downing, *Euthanasia and The Right to Die* (Nash, 1969).

[9]Marvin Kohl, *The Morality of Killing* (Atlantic Highlands, NJ: Humanities Press, 1974); *Beneficent Euthanasia* (Buffalo: Prometheus, 1975).

[10]O. Ruth Russell, *Freedom To Die* (New York: Human Sciences Press,
1975).

[11]See, for example, James Rachels, "Killing and Starving to Death," *Philosophy* 54, 1979, 159–171. This article contains Rachels' most recent and
most sophisticated version of the argument. Rachels' original argument on the
subject, "Euthanasia, Killing and Letting Die" was published in John Ladd's
excellent anthology *Ethical Issues Relating to Life and Death* (Oxford: Oxford
University Press, 1979), 146–163. A condensed version appeared in the *New
England Journal of Medicine* 292, January 1974, 78–80. The latter version
has been much anthologized and is available in Bonnie Steinbock's *Killing
and Letting Die* (Englewood Cliffs, NJ: Prentice Hall, 1980), and *Ethical Issues in Death and Dying,* edited by Tom Beauchamp and Seymour Perlin and
also published by Prentice Hall, 1978. Singer's work is in his *Practical Ethics*
(Cambridge: Cambridge University Press, 1979). Tooley's argument is in
Steinbock, pp. 56–62.

[12]Philippa Foot takes this line in "The Problem of Abortion and the Doctrine of Double Effect," *Oxford Review* 5, 1977, and in "Euthanasia", *Philosophy and Public Affairs* 6, 1977, 85–112. Foot's work is explored at some length by Nancy Davis in the Steinbock anthology, pp. 172–214, and by Bruce Russell, "The Relative Strictness of Negative and Positive Rights," *American Philosophical Quarterly* 14, 2 April 1977, (also in Steinbock), "Still A Live Issue," *Philosophy and Public Affairs* 7, 1978, 278–281 and "The Presumption of Taking Life as Compared to That Against Failing to Save," *Journal of Medicine and Philosophy* 4, 3, 1979. Russell's interlocuter and antagonist on these issues is Richard Trammell, "Saving Life and Taking Life," *Journal of Philosophy* LXXII, 5, March 13, 1975, 131–137, "Tooley's Moral Symmetry Principle," *Philosophy and Public Affairs* 5, 1976, 305–313, "The Non-Equivalence of Saving Life and Taking Life" and "The Presumption Against Taking Life," both in the *Journal of Medicine and Philosophy* 4, 1979 and 3, 1978, pp. 53–67, respectively. See also Tooley's article in Ladd, pp. 63–93 and Judith Thomson's "Killing, Letting Die, and The Trolley Problem," *The Monist* 59, 1976, 204–217.

[13]Tom Beauchamp, "A Reply to Rachels On Active and Passive Euthanasia," in Beauchamp and Perlin, pp. 246–259.

[14]Bonnie Steinbock, "The Intentional Termination of Life," *Ethics in Science and Medicine* 6, 1, 1979, 59–64, also in Steinbock, pp. 69–77.

[15]In "Killing and Starving to Death," *Philosophy* 54, 1979, pp. 165–166, Rachels develops some formally sufficient conditions for "moral equivalence". Michael Tooley, who also used this argument, presents a different theory of moral equivalence in "An Irrelevant Consideration: Killing Versus Letting Die" in Steinbock, pp. 56–62.

[16]Peter Singer, *Practical Ethics* (Cambridge: Cambridge University Press, 1979), 152–153.

[17]Beauchamp and Walters, p. 309.

[18]John Freeman "Ethics and The Decision Making Process for Defective Children," in David Roy *Medical Wisdom and Ethics in the Treatment of Severly Defective Newborn and Young Children,* (Eden Press, 1978) p. 32.

[19]Roy, pp. 28–29.

[20]"Euthanasia, Killing and Letting Die", Ladd, p. 150.

[21]M. T. Rabkin et al., "Orders Not to Resuscitate," *New England Journal of Medicine* 259, 1976, pp. 365, 366.

[22]G. Tagge et al., "Relationship of Critical Care to Prognosis in Critically Ill Patients," *Critical Care Medicine* 2, 2, March-April 1974, 61–63. See also M. Kirchner, "How Far to Go Prolonging Life: One Hospital's System," *Medical Economics* 1976, 12 July 1976, 69–75. See also G. Tagge, "Critical Care: A Classification System for ICU Patients," *The Hospital Medical Staff* 5, 2, February 1976, 1–6. Variants of this protocol have been published by Massachusetts General Hospital (Boston) and by Presbyterian University Hospital (Pittsburgh). See, Clinical Care Committee, Massachusetts General Hospital, "Optimum Care for Hopelessly Ill Patients," *New England Journal of Medicine* 295, 7, August 12, 1976,

362–364; Ake Grenvik et al., "Cessation of Therapy in Terminal Illness and Brain Death," *Critical Care Medicine* 6, 4, July–August 1978, 284–291.

²³As quoted by Rachels in Ladd, p. 149, Steinbock, p. 63, *New England Journal*, 292, January 1975, p. 78.

²⁴JAMA Aug 1, 1980, 244, 5; p. 506.

²⁵Steinbock, p. 66–67.

²⁶Ake Grenvik et al., "Cessation of Therapy in Terminal Illness and Brain Death, *Critical Care Medicine* 6, 4, July–August 1978, pp. 284–291.

²⁷These figures are projections from the data in the US Department of Health Education and Welfare's publication *Hospital Health US,* 1979. In 1977, 36.8 million discharges excluding newborns from all *short* stay hospitals (p. 129) and 1,303,100 were in nursing homes (p. 131). Figures for long-term stay hospitals, and psychiatric hospitals were not available. If all excluded groups are counted, it is reasonable to assume that well over forty million people were hospitalized for some period of time in the US in 1977. According to the Statistical Abstracts the estimated population of the US in 1977 was 216,900,000. At a conservative estimate, therefore, a North American population of 300,000,000 would generate 50,000,000 hospitalized in a given year.

²⁸*Practical Ethics,* 144–145.

²⁹*New England Journal of Medicine* 292, 16, April 17, 1975, pp. 863–864. Rachels replies on 866, 867. By the way, Singer cites this correspondence in *Practical Ethics,* p. 228.

³⁰*Ibid.,* 866, 867.

³¹Stuart Alsop, "The Right To Die With Dignity," *Good Housekeeping* 179, 2, p. 69, 130, 132, August 1974, quote from p. 132.

³²"Terminal Care: Issues and Alternatives," Ryder and D. Ross, *Public Health Reports* 92, 1, January-February, 1977. See also C. Saunders, "The Challenge of Terminal Care" in Symington and Carter, *Scientific Foundations of Oncology* (London: Heinemann, 1975). See also *The Hospice Movement* by S. Stoddard, (New York: Stein and Day, 1978).

³³*On Dying Well,* Church Information Office, London, 1975.

³⁴*Ibid.,* p. 47.

³⁵Jan. 14, 1982, 306, 2 pp. 98–99.

³⁶David Sudnow, *Passing On,* (Englewood Cliffs, NJ: Prentice Hall, 1967), p. 89.

³⁷*Ibid.,* p. 91.

³⁸*Ibid.,* p. 83.

³⁹*Ibid.,* pp. 82–83.

The Sanctity of Life

James Rachels

Two Traditions

The doctrine of the sanctity of life is associated with two great traditions of thought, one eastern and the other western. The eastern tradition is most perfectly realized in Jainism, a religion of India that is older than recorded history. The Jains believe that all life is sacred, with no sharp distinction drawn between animal life and human life. The Jain monk's devotion to this precept is awesome:

> The monk must strain his water before drinking; he wears a gauze mask over his mouth to prevent the unintentional inhalation of innocent insects; the monk is required to sweep the ground before him as he goes, so living beings are not crushed by his footsteps; and always he treads softly, for the very atoms underfoot harbour minute life-monads.[1]

The monk's discipline cannot be matched by the ordinary Jain, who as a practical matter cannot take up the gauze mask and the broom. Nevertheless, the monk's practice stands as an ideal, and the others do what they can.

The example of the Jains is a little intimidating to westerners who like to think that they also believe in the sanctity of life. At least on the surface, the Jains appear to be more consistent. We abhor the killing of humans, but, unlike the Jains, we think nothing of crushing bugs beneath our feet. Yet, if life is sacred, why is the life of a bug not sacred? A bug is a living thing just as surely as we humans are living things.

A few westerners have accepted this conclusion and have embraced the eastern view. The most notable is Albert Schweitzer, the doctor, scholar, musician, missionary, and Nobel peace prize winner who died in 1965. This extraordinary man preached an ethic of "reverence for life" that would protect not only all animals, but plants as well. The life he recommends is very much like that endorsed by the Jains:

> A man is really ethical [says Schweitzer] only when he obeys the constraint laid on him to help all life which he is able to succour, and when he goes out of his way to avoid injuring anything living. He does not ask how far this or that life deserves sympathy as valuable in itself, nor how far it is capable of feeling. To him life as such is sacred. He shatters no ice crystal that sparkles in the sun, tears no leaf from its tree, breaks off no flower, and is careful not to crush any insect as he walks. If he works by lamplight on a summer evening, he prefers to keep his window shut and to breathe stifling air, rather than to see insect after insect fall on his table with singed and sinking wings.[2]

There is something undeniably attractive about this view of moral life, especially when we remember Schweitzer's own example. He devoted many years to working among impoverished people—this is what earned him the Nobel peace prize—and it is easy to imagine him breathing the stifling air at his mission rather than opening his window to the insects. It is tempting to conclude, uncritically, that the moral view he articulates, one that his own life exemplifies, is a higher and nobler ideal than most of us are willing to embrace. Yet a more sceptical attitude might be in order. We might wonder whether, from an objective point of view, this moral position is correct. Are there any compelling reasons why a sensible person should accept it?

Schweitzer's reasons were different from the Jains' reasons. The Jains believe that each human soul passes through an immense number of reincarnations, all the while striving for release. Upon its liberation, the soul ascends to a level of existence higher than the heavens, where it is at peace for eternity. Morality is viewed by them as a means of progressing toward liberation. Respect for life is a part of the selfless spiritual discipline that leads ultimately to nirvana. Obviously, this will provide a powerful reason for those who have grown up within that culture. But for those who do not share the Jain world view, all this will seem merely bizarre.

Schweitzer relied on a very different sort of argument. Each of us, he said, has a "will-to-live." We recognize in ourselves "a yearning for more life"; we value our lives and we do not want to die. He thought that if we attend sympathetically to other living things, we will recognize the same will-to-live in them. Apparently this will-to-live does not depend on a conscious desire, for plants are said to have it too. Be that as it may, Schweitzer thought that consistency requires us to show respect for other life, if we wish respect to be shown for our own. He concludes that

> Ethics consists in this, that I experience the necessity of practicing the same reverence for life toward all will-to-live, as toward my own.

Albert Schweitzer was, however, an exception. The dominant tradition in the west is strikingly different from the eastern way of thinking. The western tradition does not hold that all life is sacred; rather, it emphasizes the difference between human and nonhuman life, and grants serious moral protection only to the former. Whereas the Jains are vegetarians, the western tradition sees nothing wrong with eating meat, or with killing animals for any number of other purposes. Imagine what a Jain would think upon reading St. Augustine, who says:

> Christ himself shows that to refrain from the killing of animals and the destroying of plants is the height of superstition, for judging that there are no common rights between us and the beasts and trees, he sent the devils into a herd of swine and with a curse withered the tree on which he found no fruit.[3]

St. Augustine was, of course, one of the most important Christian thinkers, and the western view of the sanctity of life is largely the product of Christian influence. Church doctrine on this subject, however, developed in an interesting way. Although they were indifferent to the lives of nonhuman animals, the early church fathers taught that it is wrong to kill human beings *for any reason whatever*. This set the church very much against the prevailing culture. The Greeks and Romans had permitted euthanasia, or suicide, as an alternative to a lingering death. They had also allowed the killing of deformed infants. The church vigorously condemned both practices. Capital punishment was also condemned by the church. Moreover, although it comes as a surprise to most modern-day Christians, the early church was resolutely opposed to killing in war. The Christian rule was that killing human beings is wrong, period.[4]

As time passed, the rule was softened. Christianity grew and the possibility arose that it might become a state religion. But it could not do so if it continued to oppose the settled policies of the state. Therefore, the church dropped its opposition to capital punishment, and said only that priests should take no part in it. The ban on war was also modified. Rather than condemning all war, the church now developed a doctrine of "the just war," and said that, if the war is just, killing is permissible.

These developments forced the church to face a problem of theory. As long as all killing was forbidden, it was clear what principle the church followed. The principle was that all killing of human beings is wrong. But after war and capital punishment were accepted, that principle could no longer be claimed. What, then, was the new principle the church affirmed? What principle would allow killing in war and as punishment, but prohibit killing in all those other circumstances?

The solution to this problem was found in the idea that neither the enemy soldier nor the criminal is "innocent." The soldier is not innocent because he unjustly threatens the lives of others, and the criminal is not innocent because he violates society's laws. On the other hand, the victims of the other types of killing, such as infanticide and euthanasia, *are* innocent. Once this was noticed, the new rule was easy to formulate. It was not to be the simple rule that killing people is always wrong. Instead, it was the more sophisticated rule that the killing of *innocent* people is always wrong.

Only a little reflection is needed to see how much we stand in this tradition. Even today, the moral feelings of most people in the west are still shaped by that great division of killings into permissible and impermissible. Most people still take it for granted that killing by soldiers is all right; and, although capital punishment is controversial, most people, at least in the United States, continue to favor it. And most people still believe that infanticide, suicide, and euthanasia are wrong. It is sobering to realize how much our moral views are a product of that ancient compromise between church and state.

At any rate, the difference between the eastern and the western traditions can be summarized very simply: the eastern doctrine is a doctrine of the sanctity of *all* life, while the western doctrine is a doctrine of the sanctity of *innocent human* life.

Despite their great influence, it is easy to find fault with these doctrines. Neither is very plausible. Three obvious objections come to mind:

First, with its emphasis on the protection of all life, the eastern tradition has difficulty in distinguishing among various *kinds* of lives. The life of a person and the life of a bug seem to be equally important. But of course there are many significant differences between persons and bugs. Humans (and, for that matter, many other species of animals as well) have mental and emotional lives far superior to that of insects. Why should this not put them in different moral categories? It is far too simple-minded to say, as Schweitzer does, that "all life is one." The differences are important.

Second, if the eastern tradition pays too much respect to nonhumans, the western tradition pays far too little. St. Augustine expressed the western attitude very well when he said that "there are no common rights between us and the beasts"; thus, we kill intelligent and sensitive animals, not only for food, but to test cosmetics, for ingredients for perfume, and simply as sport. Surely there ought to be some mean between the extremes represented by the two traditions.

Third, although it seems a shocking thing to say, the western tradition places *too much* value on human life.[5] There are times when the protection of human life has no point, and the western tradition has difficulty acknowledging this. The noble ideal of "protecting human life" is invoked even when the life involved does its subject no good and even when it is not wanted. Babies that are hopelessly deformed, and will never mature into children, may nevertheless be kept alive at great cost. Euthanasia for persons dying of horrible diseases is illegal. The babies and the hopelessly ill are "innocent," of course, but that is sadly irrelevant. St. Augustine called respect for animal life "the height of superstition"; in these cases, it is respect for human life that seems to have degenerated into a mere superstition.

In my opinion, all these objections are sound, and both the eastern and the western doctrines ought to be rejected. But these objections are also, in a certain sense, superficial. At a deeper level, there is one fundamental mistake that underlies both doctrines. This is somewhat surprising, considering how different they are. Nevertheless, this one basic mistake explains why both are inadequate guides for conduct. I want to explain what that mistake is, and at the same time show what a better understanding of the sanctity of life would be like.

Two Concepts of Life

We use the word "life" in two ways. On the one hand, when we speak of "life," we may be referring to living things, to things that are *alive*. To be alive is to be a functioning biological organism. Here the contrast is with things that are dead, or with things that are neither alive nor dead, such as rocks. Not only people, but chimpanzees and bugs, and even trees and bushes, are living things.

On the other hand, when we speak of "life," we may have in mind a very different sort of concept, one that belongs more to biography than to biology. Human beings not only are alive; they *have lives* as well.[6]

It is not easy to mark this distinction in English, since one word does double duty. The distinction would be easier to spot if the language provided two words. Nonetheless, the difference between being alive and having a life is obvious enough. If we were to describe

Bobby Fischer as a living being, we could say that he is an animal of the species *Homo sapiens,* that he has a heart and liver and brain and blood and kidneys that function together in a certain way, and so forth. If we were to describe his *life,* however, we would mention an entirely different set of things. The facts of a person's life are facts about his history and character. Fischer was born in 1943 and grew up in New York City; he learned to play chess at age six and devoted himself single-mindedly to the game thereafter. He became the United States Champion at 14 and dropped out of high school; he won the world championship in 1972 and has been a recluse ever since. He has always suspected Russian players of trying to cheat him, and more generally, he trusts almost no one. He was involved with an off-beat religious cult in California, but that ended in a public dispute. These are some of the facts of his life.

If the concept of life is ambiguous in this way, then so is the concept of the sanctity of life. The doctrine of the sanctity of life can be understood as placing value on things that are alive. But it can also be understood as placing value on *lives* and on the interests that some creatures, including ourselves, have in virtue of the fact that they are the subjects of lives. Very different moral views will result, depending on which interpretation one chooses.

The distinction between being alive and having a life has not been carefully observed in either the eastern tradition or the western tradition. But the emphasis in both has been on the protection of living things. In the eastern tradition, this is abundantly clear. The Jain monk sweeps the ground before him, so that insects are not crushed beneath his feet. But insects, while they are doubtlessly alive, do not have lives. They are too simple. They do not have the mental wherewithal to have plans, hopes, or aspirations. They cannot regret their pasts, or look forward to their futures. In short, they lack a psychology. Plants, toward which Schweitzer recommends reverence, are not even conscious. An ethic of reverence for *lives* would find little here with which to concern itself.

The western tradition is a bit more complicated on this point. The doctrine of the sanctity of life is taken to protect only humans, and humans do have lives, so it might appear that the western doctrine is concerned with lives. However, the western doctrine is more concerned with the fact of being human than with the fact that humans have lives. The sanctity of life is interpreted as applying to *all* humans, even those that do not have lives, such as hopelessly deformed babies who cannot mature into children, or persons in irreversible comas. So, after limiting itself to humans, the western doctrine pays no attention to the difference between humans who have lives and those who do not.

This is the fundamental mistake that I think underlies both the eastern and the western traditions. The sanctity of life ought to be interpreted as protecting lives in the biographical sense, and not merely life in the biological sense. There is a simple, but I think conclusive, argument for this. From the point of view of the living individual, there is nothing important about being alive except that it enables one to have a life. In the absence of a conscious life, it is of no consequence to the subject himself whether he lives or dies. Imagine that you are given the choice between dying today and lapsing into a dreamless coma, from which you will never awaken, and dying ten years from now. You might prefer the former because you find the prospect of a vegetable existence undignified. But in the most important sense, the choice is indifferent. In either case, your *life* will end today, and without that, the mere persistence of your body has no importance. Therefore, insofar as we are concerned to protect the interests of the individuals whose welfare is at stake, we should be concerned primarily with lives and only secondarily with life. This does not mean, of course, that being alive is unimportant. In the great majority of cases it is very important, for you cannot have a life if you are not alive. But the importance of being alive is only derivative from the more fundamental concern for lives.

We have, then, if I may speak somewhat grandly, a new understanding of the sanctity of life. The practical implications are different from the implications of both the traditional eastern and western doctrines, and in each case, I believe the new understanding provides a better ethic. A few examples will illustrate why this is so, and will make the nature of my suggestion clearer.

"Human Vegetables"

A man named Repouille, living in California, killed his 13-year-old son with chloroform. The boy had suffered a brain injury at birth and was blind, mute, deformed in all four limbs, and virtually mindless. With no control over his bladder or bowels, he had lain in a small crib since birth. Repouille's other four children were normal, and he was a normally good father to them.

Repouille was tried for manslaughter in the first degree, but the jury, obviously sympathetic with him, brought in a verdict of second-degree manslaughter. (Technically, he was guilty of first-degree murder.) The jury also recommended that the judge exercise "utmost clemency" in sentencing. The judge agreed with this sentiment, and placed Repouille on probation. He never went to jail.

All this happened more than forty years ago, in 1939, but it could have happened yesterday. Neither the legal situation, nor the moral climate that nurtured it, have changed. Such killings still occur, the law still officially condemns them, and the courts are still reluctant to impose heavy penalties.

The law condemns such killings because it reflects the western tradition concerning the sanctity of life. The western tradition does not distinguish between the life of a normal, healthy person and the ''life'' of Repouille's son lying in the crib: both are sacred. So the law does not officially recognize any difference between killing a normal, healthy person and killing Repouille's son: both are murder. Yet the human beings who must deal with such killings—the judges and juries—are obviously uncomfortable with this attitude.

An ethic that emphasized the protection of lives and not mere biological life would say that the judges and juries are right. Repouille's son was alive, but the tragic brain injury prevented him from ever having a life. Thus, on the view I am proposing, it is a mistake to speak of ''the sanctity of life'' in this instance.

This is easy enough to say, but difficult to apply in such a heartbreaking case. We feel a natural sense of identification with other humans, probably for deep biological reasons, no matter how unfortunate they are. But we should ask of Repouille's son whether his ''life'' did *him* any good. Did it have any value from *his* point of view? And immediately we face the crucial problem that he did not have a point of view, for he was not the subject of a life.

However, he was presumably conscious, at least to the extent of being able to experience pleasure and pain. This may lead some people to reject the view I am suggesting. But the capacity to experience simple pleasures and pains is not much, if it is the only capacity one has. The sensations experienced will not arise from any human activities or projects; they will not be connected with any coherent view of the world. Nevertheless, one could maintain that the bare capacity to suffer and enjoy is enough to make the protection of life morally important. One could say that, *if* one were willing to extend equal moral protection to mice, fish, and toads, who, after all, can also experience pleasure and pain. However, I think the ethics of killing involves fundamentally different issues than the ethics of causing pain, and the two ought to be kept separate. The moral rule against causing pain applies to all creatures that are capable of suffering; but something more is required to come under the protection of the rule against killing. On the view that I am suggesting, the rule against killing applies only to creatures that have lives.

The case of Repouille's son is similar in some ways to the celebrated case of Karen Ann Quinlan, but there is an important difference.

Whereas Repouille's son never had a life, Karen Quinlan did have one prior to entering the coma. She was a happy young woman, a devout Catholic, with loving parents. As this is written, she is still alive, although unconscious, deformed permanently into a fetal position. But her life in the biographical sense ended when she entered the coma, and so the continuing efforts to maintain her alive can signify only a concern for her as a living thing. On the understanding of the sanctity of life that I am proposing, those efforts are sadly pointless.

Defective Infants

In the October 25, 1973 issue of *The New England Journal of Medicine* two articles appeared that caused great public controversy.[7] In one, Drs. Raymond Duff and A. G. M. Campbell described how they had let 43 defective babies die in the Yale-New Haven Hospital's Special Care Nursery. In the other, Dr. Anthony Shaw of the University of Virginia Medical Center discussed his own practice regarding such cases. Like Duff and Campbell, Shaw said that he had sometimes allowed defective babies to die when the parents refused permission for needed surgery.

Physicians were aware of such practices; the general public was not. Allowing defective infants to die had become common, though unpublicized, as medical techniques for keeping babies alive had become more sophisticated. The doctors argued that the new techniques should not be used simply because they were available; decisions were required in individual cases about whether they *ought* to be used. Shaw observes:

> Each year it becomes possible to remove yet another type of malformation from the "unsalvageable" category. All pediatric surgeons, including myself, have "triumphs"—infants who, if they had been born 25 or even five years ago, would not have been salvageable . . . [But] how about the infant whose gastrointestinal tract has been removed after volvulus and infarction? Although none of us regard the insertion of a central venous catheter as a "heroic" procedure, is it right to insert a "lifeline" to feed this baby in the light of our present technology, which can support him, tethered to an infusion pump, for a maximum of one year and some months?

Among the 43 infants allowed to die at the Yale-New Haven Hospital were 15 with multiple anomalies, eight with trisomy, eight with cardiopulmonary disease, seven with meningomyelocele, three with other central nervous system disorders, and two with short-bowel syndrome. Duff and Campbell argue that nontreatment was justified because in each case, "prognosis for meaningful life was extremely poor or hopeless."

Does the nontreatment of these infants violate the sanctity of life? Duff and Campbell comment, "If this is one of the consequences of the sanctity-of-life ethic, perhaps our formulation of the principle should be revised." Such a revision is, of course, exactly what I have proposed. On my principle, the question to be asked is whether these babies have any prospect of a life in the biographical sense.

The infant described by Dr. Shaw obviously has no such prospect. The baby could be kept alive by a central venous catheter for less than two years; then it would die. It would be conscious during that time—it might be aware of having its diaper changed, and of being hungry, and of the pain associated with its medical condition—but that is about all. It is no better off than Repouille's son; it will never have a life, and so the doctrine of the sanctity of life will have no application.

Unfortunately, however, my principle does not provide a straightforward recommendation concerning many of these babies. The case described by Dr. Shaw is clear-cut: the baby has no prospect of a life, and that is all there is to it. But in other cases, it may not be clear whether the infant has such a prospect. An infant with Down's syndrome (mongolism) may develop to an impressive degree, or it may not. The problem is that we cannot tell how severe the retardation is until we observe the child's progress, or lack of it. Therefore we cannot know when the baby is born how much of a life it has in store. Without such information, we cannot judge to what extent the principle of the sanctity of life should be applied. This is not a deficiency of the principle, but only reflects our lack of factual knowledge.

Nonetheless, there are some cases in which we *do* know that an infant has no prospect of a life, and in those cases the principle I am proposing would have us stop worrying about the sanctity of life.

The Terminally Ill

Skin cancer had riddled the tortured body of Matthew Donnelly. A physicist, he had done research for the past thirty years on the use of X-rays. He had lost part of his jaw, his upper lip, his nose, and his left hand. Growths had been removed from his right arm and two fingers from his right hand. He was left blind, slowly deteriorating, and in agony of body and soul. The pain was constant; at its worst, he could be seen lying in bed with teeth clenched and beads of perspiration standing out on his forehead.

Nothing could be done except continued surgery and analgesia.
The physicians estimated that he had about a year to live.[8]

Mr. Donnelly begged his brother to shoot him, and he did.

Euthanasia is, of course, illegal, at least in the United States. But
is it immoral? There is one clear argument in favor of it, namely, that it
puts an end to suffering. But in the western tradition this is not considered sufficient reason to set aside the prohibition on killing. The dying
person is "innocent," and therefore comes under the rule against killing the innocent. Joseph Sullivan, a Catholic bishop and prominent op-
ponent of euthanasia, stated the classic objection when he wrote: "It is
never lawful for man on his own authority to kill the innocent directly.
If this thesis is morally sound, it follows that mercy killing is never
permissible."[9]

An ethic that is concerned with lives would lead to a different conclusion. It would have us ask what effect killing Mr. Donnelly would
have on his biographical life. In the case of Karen Quinlan, the answer
is that killing her would have no effect whatever on her life: her life is
already over; it ended when she entered the coma. Matthew Donnelly's
life, however, was still in progress when his brother shot him. So a
superifical answer would be that killing him destroyed his life.

Less superficially, we should compare the life Donnelly would
have had if he were not killed with the life he would have had if he
were killed. Then we can see the difference euthanasia makes. The two
lives are exactly the same, except that one is a year longer. The extra
year, however, is not spent on any project or activity that Donnelly
thinks worthwhile; it is spent lying in bed, blind, deformed, teeth
clenched, in constant pain. Is his life better with or without this extra
year? Surely, if Donnelly believed that this extra year of being alive
would add nothing of value to his life, that is a reasonable judgment.
The sanctity of life, if it were interpreted as protecting lives and the
interests that people have in virtue of the fact that they are the subjects
of lives, would offer no objection to euthanasia in this case.

As I have already remarked, the ancient Greeks and Romans
would have found nothing controversial in this. In fact, many of them
would go even farther than I am willing to go. The Stoic philosophers,
for example, thought a person is morally free to end his life *whenever*
he finds it no longer worth living. Epictetus compared life to a smoky
room: "If the room is smoky, if only moderately, I will stay; if there is
too much smoke I will go. Remember this, keep a firm hold on it, the
door is always open."[10] On this view, it seems that no one has a duty
to preserve his or her own life; the lives of others may be sacred, but
not one's own. The interpretation of the sanctity of life I am proposing

JAMES RACHELS

would not agree with this. Most suicides are tragically short-sighted; they occur when people are temporarily blinded by the smoke and cannot see that in fact their life-prospects are not so bad. Concern for lives must include a concern for one's own life, which is no less important than anyone else's. Morally speaking, the door is not always open. It is open only when being alive no longer adds anything to one's life.

Nonhuman Animals

The most striking difference between the eastern and western traditions is in their attitudes toward nonhuman animals. One tradition grants all of them the full protection of the doctrine of the sanctity of life; the other gives them virtually no protection at all. As I have already suggested, we need to find some reasonable middle ground.

Do animals have lives? Some of them clearly do not. Having a life requires some fairly sophisticated mental capacities, which simple animals do not have. Consider, however, the rhesus monkey, which is not a simple animal. Rhesus monkeys live together in social groups; they communicate with one another; they engage in complicated activities; they have highly individualized personalities. They are so much like humans that they are favorite research animals for psychologists seeking to learn what makes us tick. One team of researchers noted that they are "much more mature intellectually than a human at birth," and that they "can indeed solve many problems similar in type to the items used in standard tests of human intelligence."[11] Although their lives are not as complicated as ours, emotionally or intellectually, there seems no doubt that they do have lives. They are not merely alive.

Looking at the entire animal kingdom, the situation seems to be this. When we consider the higher animals—the mammals with which we are most familiar—we find that they have lives. But the farther down the phylogenetic scale we go, the less certain this seems, until we reach the clams and snails and bugs, which surely do not have lives in any but a metaphorical sense. On my version of the sanctity of life, then, the higher animals ought to have serious moral protection—to kill a rhesus monkey or a chimpanzee might not require quite as strong a justification as killing a person, but it is nonetheless a serious matter. The protection becomes weaker, however, as we consider progressively simpler animals, until we reach the clams and snails and bugs whose "lives" do not count for much at all. Thus, we have found a reasonable middle ground between the two extremes represented by the eastern and western traditions.[12]

These are enough examples to illustrate the difference between the new understanding of the sanctity of life and those offered by the two major traditions. I believe that observing the difference between being alive and having a life provides a simpler, more systematic, and morally more satisfactory way of approaching the subject than is provided by either tradition.

In presenting this new approach I have retained the traditional terminology, and that requires a brief comment. The word "sanctity" has religious associations; but the doctrine that I have proposed is not a religious doctrine. It is an ethical idea that does not require a religious justification. It may be accepted or rejected on its own merits, either by religious people or by nonreligious people. However, this is not all that different from the traditional Christian position concerning ethics, which holds that moral truths are truths of reason binding on all people, regardless of their religious convictions or lack of them. That is all I wish to claim for this new understanding of the "sanctity" of life.

Notes and References

[1]Ninian Smart, *The Religious Experience of Mankind* (London: Fontana Library, 1971), p. 106.

[2]Albert Schweitzer, *Civilization and Ethics,* translated by John Naish; reprinted in Tom Regan and Peter Singer (eds.) *Animals Rights and Human Obligations,* (Englewood Cliffs, NJ: Prentice-Hall, 1976), p. 134.

[3]St. Augustine, *The Catholic and Manichaean Ways of Life,* translated by D. A. Gallagher and I. J. Gallagher (Boston: Catholic University Press, 1966), p. 102.

[4]There were some small exceptions made, which for present purposes we may ignore. For example, a virgin was allowed to commit suicide to avoid being raped. Virginity was very highly prized. This dispensation remained in effect until St. Augustine argued that chastity is more a matter of the mind than the body, but there is no record of how many women took advantage of it.

[5]On this point, see Peter Singer, "Unsanctifying Human Life," in John Ladd (ed.), *Ethical Issues Relating to Life and Death* (New York: Oxford University Press, 1979), pp. 41–61.

[6]I have learned much about this distinction from William Ruddick in conversations over a period of years. He makes use of it in his essay "Parents and Life Prospects," in Onora O'Neill and William Ruddick (eds.) *Having Children* (New York: Oxford University Press, 1979), pp. 123–137.

[7]Anthony Shaw, "Dilemmas of 'Informed Consent' in Children," *The New England Journal of Medicine* 289 (1973), 885–890; and Raymond S.

Duff and A. G. M. Campbell, "Moral and Ethical Dilemmas in the Special-Care Nursery," *The New England Journal of Medicine* 289 (1973), 890–894.

[8]Robert M. Veatch, *Case Studies in Medical Ethics* (Cambridge, MA: Harvard University Press, 1977), p. 328.

[9]Joseph Sullivan, "The Immorality of Euthanasia," in Marvin Kohl (ed.), *Beneficent Euthanasia* (Buffalo, NY: Prometheus Books, 1975), 12.

[10]Epictetus, *Dissertations*, I, IX, 16.

[11]H. F. and M. K. Harlow, *Lessons from Animal Behavior for the Clinician* (London: National Spastics Society, 1962), ch. 5.

[12]For more on the question of whether animals have lives, and its moral implications, see J. Rachels, "Do Animals Have a Right to Life?" in Harlan B. Miller and William H. Williams (eds.), *Ethics and Animals* (Clifton, NJ: The Humana Press, 1983).

Section II
Surrogate Gestation

Introduction

In "Surrogate Gestation: Law and Morality" Theodore Benditt, who is both an attorney and a philosopher, begins by focusing on the legality of the practice of laboratory-assisted surrogate gestation. In the first instance he concludes, after a detailed examination, that there is no constitutional prohibition of the practice even though the plausible idea that it is a fundamental right is hardly conclusive. Secondly, after comparing the legality of surrogate gestation with similar practices that have clearer legal standing, he concludes that, though there are certain legal problems attending the practice of surrogate gestation, there is no strong reason for thinking that these problems could not be resolved in favor of the practice.

Professor Benditt then goes on to consider the moral issues involved in surrogate gestation. After examining a number of pointed arguments against the morality of the practice, he concludes that they are not forceful objections. All in all, Benditt finds nothing either legally or morally objectionable with the practice.

In "Surrogate Motherhood: The Ethical Implications," Lisa H. Newton explores the many ethical issues involved in laboratory assisted surrogate motherhood. Questions of efficiency aside, she asks how research into embryonic transplants can be justified in the presence of world poverty, starvation, overpopulation, and rampant sickness. Should the public be paying for this sort of research for the gratification of a few couples when childlessness is by no means a life-threatening condition? Like Benditt, she also considers the legal issues involved and focuses on the further moral issues bearing on the possible exploitation of both the mother and the embryo. She concludes, like Benditt, that neither set of considerations generates any serious reason for thinking the practice is either illegal or immoral. She also reviews objections from the area of religion and fails to see any real force in them.

In the end, while Professor Newton professedly offers only halting conclusions, or way stations on the road to the conclusions, the reader may well conclude that the only apparent objection to the practice has to do with whether society should be forced to pay for the research involved when it can plausibly be argued that society has other more pressing needs that should be satisfied first. To what degree this last consideration bears on the morality or the immorality of the otherwise morally and legally acceptable practice of laboratory assisted surrogate motherhood may well be the focus of future discussions.

Surrogate Gestation

Law and Morality

Theodore M. Benditt

Introduction

Alice and Joe are married. They would like to have a child, but cannot, because Alice is unable to bear a child. Alice has a sister, however, who can conceive and is willing to be artifically inseminated with Joe's sperm, to carry and deliver the child, and then to turn it over to Alice and Joe. Alice's sister will give up whatever legal rights she may have with respect to the child, and Alice and Joe will take whatever steps are required to become its parents.

Alice's sister is what is called a surrogate mother, though 'surrogate baby-bearer' or 'surrogate gestator' are better terms, since being a mother suggests more than merely bearing a child. This article is concerned with the legal and moral status of the arrangement described above, and in particular with such questions as whether Alice and Joe have any Constitutionally protected rights in the matter, whether there are any serious moral difficulties with the arrangement, whether there should be laws against such arrangements, and whether legal difficulties arising from such arrangements, such as the child's legitimacy, access to information about the circumstances of its birth, and the rights and responsibilities of the surrogate, can be satisfactorily dealt with. These are particularly important questions at this time because it is estimated that as many as 15% of married couples are unable to have children in the usual way, and because medical technology is now able to offer alternatives.

There are variations of the arrangement outlined above, some of which raise additional questions that might make a difference in what

47

we think about it. The following three variations seem especially important. (1) Alice and Joe cannot find anyone who is willing to volunteer to help them, so they find someone who is willing to be a surrogate for a fee—say $5,000 in addition to expenses. (2) Alice is not sterile, and so an ovum is removed from her by the new *in vitro* fertilization (IVF) techniques, is impregnated with Joe's sperm, and the resulting blastocyst is implanted into a surrogate. (3) Alice and Joe could have a child in the ordinary way, but they do not wish Alice to go through the inconvenience (and risk) of child bearing, and resort instead to a surrogate, in the way described in (2).

Let us suppose, to begin with, that a state legislature wishes to prohibit surrogate gestation (hereafter SG). One possibility would be for a legislature to pass a statute making it a criminal offense to procure a woman to be a surrogate or to enter into any sort of agreement involving SG. No legislature has passed such a law, though other legislation, having to do with various aspects of baby-bearing and parent–child relationships, has been applied in ways that hamper SG arrangements.

Constitutionality

The threshold question is whether, and/or to what extent, legislative restriction or regulation of SG is Constitutional. Legislation limiting an individual's behavior must, of course, pass Constitutional muster, but there are two different standards governing the permissibility of such legislation, depending on whether fundamental rights of the parties are involved. If a person has a fundamental right with respect to something, he or she may not be prevented from doing it. That does not mean that it may not be regulated, but it does mean that it may be regulated only if there are particularly important public or governmental interests at stake. The role of the courts in such cases is to be protective of the fundamental rights, and to look very closely at the state's claim that regulation is needed despite the infringement of fundamental rights. On the other hand, if no fundamental right is involved, due process imposes a far less stringent standard on legislation—it need only be reasonable, that is, rationally related to some element of public health, welfare, or safety, and it is inevitable that great weight will be given by the courts to a legislative determination that regulation is needed in order to protect certain public or governmental interests.

There are three fundamental rights recognized by the Supreme Court in recent years that might be at stake in matters of SG. One is the right to marital privacy recognized in *Griswold v. Connecticut.*[1] In this case, the defendants, a doctor (Buxton) and the Executive Director of Planned Parenthood of Connecticut (Griswold), had been charged with violating a Connecticut statute that forbade the giving of information, instruction, or medical advice to married couples regarding the means of preventing conception. The Court held that though it is not explicitly mentioned in the Constitution, within the penumbra of a number of the guarantees of the Bill of Rights there is a protected zone of individual privacy. The Court said that this zone includes the right to marital privacy and the right to marry and raise a family, and that the relationships that the state of Connecticut sought to regulate are included within those rights. In addition to the right to marital privacy announced in *Griswold,* there are two other fundamental rights recognized by the Supreme Court that might be called upon to protect SG arrangements. One is the right to procreate recognized in *Skinner v. Oklahoma,* which involved a law authorizing the sterilization of habitual offenders. The Court held that law violative of the fourteenth amendment, saying that an individual sterilized under the statute would be "deprived of a basic liberty."[2] More recently, in *Eisenstadt v. Baird* and *Carey v. Population Services International,* the Supreme Court, in striking down statutes restricting the distribution of contraceptives, has recognized an individual's fundamental right to decide whether to bear or beget a child. In *Eisenstadt,* the Court referred to "the right of the *individual . . .* to be free from unwanted governmental intrusion into matters so fundamentally affecting a person as the decision whether to bear or beget a child."[3] A few years later, in *Carey,* the Court said "The Constitution protects individual decisions in matters of childbearing from unjustified intrusion by the State."[4]

It is not hard to see how one might fashion an argument for fundamental rights in SG on the basis of the fundamental rights identified above—namely, the right to marital privacy, the right to procreate, and the right to decide whether to bear or beget. Given these rights, one might argue, the law cannot deprive one of his or her means of procreation, nor can the state effectively foreclose one's choice as to whether to have a child. Yet this is what a ban on SG would do: for Alice and Joe it would deprive them of the only means available of having a child genetically linked to one of them, and would thus effectively take out of their hands the decision whether or not to have a child.

But though the argument for a fundamental right is plausible, it is hardly conclusive. There is a vast difference between, on the one hand, the state's depriving a person of the means of procreation by taking away his or her fertility, and depriving a person of the means of procreation by disallowing the impregnation or implantation of a surrogate. On this issue it would undoubtedly be argued, quite plausibly and probably successfully, that the fundamental right extends only to the usual modes of procreation, and that there is no fundamental right to procreate by whatever method one wishes, even if it is the only way possible.

How far from 'normalcy' a fundamental right to procreate might extend can be tested by considering the following. Suppose Alice can bear a child, but is unable to conceive. An ovum is removed from her, impregnated *in vitro* with Joe's sperm, and later the blastocyst is implanted into Alice's uterus. Here we have a husband and wife each of whom will be a biological parent of the child, the wife will carry the offspring, and the procedure is necessary if these elements of 'normal' procreation are to be realized. The only difference from the usual is the means of bringing about the pregnancy. This is surely the strongest nonstandard case for the recognition of a fundamental right. In contrast with this, the case for a fundamental right where a surrogate is used is considerably weaker. This is not to say, however, that no case can be made for it, and Alice's and Joe's situation is very appealing. Yet it is pretty far from the usual way of procreating, and my guess is that a fundamental right to procreate in this way would not be recognized.

The situation seems to be similar with respect to the fundamental right to decide whether to bear or beget. It is true that a law prohibiting SG would effectively foreclose Alice's and Joe's decision. But in the cases announcing this fundamental right, the point seems to have been to enable couples better to control their lives by controlling the consequences of sexual intercourse. To be sure, Alice and Joe do want to take control of their lives by *not* having their parental possibilities determined for them by biological happenstance, and from that point of view their argument is quite plausible and their case again appealing. Despite this, though, I doubt that the fundamental right in question would be extended to impregnating a woman not a party to the marriage.

I will assume, for the remainder of this discussion, that there is no fundamental right to SG. That only means, however, that if a state passes a law prohibiting or regulating SG, judicial scrutiny is limited to whether there is a rational basis for the legislation. I assume that such legislation would easily pass this test, so that if a state passes such laws there will be no Constitutional barriers, and Alice and Joe may be prevented from procreating. But there is still a lot of room for argument— about whether any state *should* prohibit SG.

Legality

Similarities to Other Practices

Artificial Insemination

There are two types of artificial insemination—homologous (AIH), in which the husband's semen is used, and heterologous (AID), in which donor semen is used. AIH has never been thought to present any moral or legal difficulties. It is simply a way of achieving what cannot be achieved in the usual way—namely, the impregnation of a woman by her husband, with the result that a married man and woman are the biological parents of an offspring which the wife carries and bears. The use of IVF in similar circumstances is regarded as unproblematic in just the same way, for it is just a way of achieving exactly the same results, which cannot be achieved in the usual way.

AID is employed when the husband is sterile, but the couple nevertheless wants a child that is theirs in the sense of being the biological offspring of one of them (the wife) and being carried by her. Two things are absent in these cases that are present in the usual mode of having a child: the husband is not the biological parent of the child, and impregnation does not take place in the usual way. SG is similar to AID in the two respects in which the latter differs from the usual way of producing babies. First of all, there is no sexual intercourse involved—impregnation of the surrogate is by artificial insemination. Second, one of the spouses—this time the woman—is not the biological parent of the offspring. The new feature that SG adds is that gestation is carried out by a woman who is not a party to the marriage and who is not going to be the one who raises the child. (Notice, by the way, that SG can occur where both partners to the marriage are biological parents of the offspring. These are the cases in which both husband and wife are fertile, but the wife cannot conceive or bear a child. Fertilization would be accomplished *in vitro*, with later implantation into the surrogate. These cases differ from the usual mode of procreation in that there is no sexual intercourse and no gestational link between wife and child.)

In some respects the case discussed parenthetically above, involving IVF, is the pure case of SG, the one that most directly presents the issues unconfused by other considerations. However, this is not the usual situation, both because IVF is a new procedure that has not been perfected, and because it is not a possibility where the wife is infertile. As a result it is necessary to consider the issue of surrogation in the context of gene donation, and thus a discussion of AID is called for.

There are some early cases that have held or implied that AID constitutes adultery. There have been no actual criminal prosecutions for adultery; the issue has arisen in divorce and support cases. At a time when divorces were harder to get, a husband might, for example, have based a suit for divorce on the ground that his wife, having had AID without his consent, is guilty of adultery. Or a woman might bring an action against her former husband for child support, and he might disclaim liability on the ground that AID constitutes adultery. An unreported lower court case in Illinois in 1945 held that AID was not adulterous,[5] but a later case in Illinois,[6] in 1954, held that it was, as did a 1963 case in New York. But the idea of adultery without a sexual act, and based instead on the doing of something that might bring about a conception that is by some lights illegitimate, seems farfetched. A Scottish case in 1958 held that adultery requires a sexual act,[7] and a California Supreme Court case in 1968 derided the idea that AID could constitute adultery, pointing out that the semen could be injected by a female doctor or by the husband himself, and that the semen donor could be far away or even dead.[8]

AID raises some important problems having to do with the status of the child—legitimacy, right to inherit, and liability of husband for child support—that are not answered merely by concluding that AID is nonadulterous. A few earlier cases held that an AID child was illegitimate. As a practical matter, though, it would be hard to prove that an AID child is not the offspring of the woman's husband. For a child born during marriage is presumed to be legitimate, and clear and convincing evidence is needed to rebut the presumption. But most men unable to father a child in the usual way are not completely sterile, and doctors often mix the husband's semen with the donor's. Further, donors are often selected who have blood type and physical characteristics similar to the husband.

Recent cases, in any event, have held the AID child legitimate, at least where the husband has consented. Both a California court in 1968 and a New York court in 1973 held that the interests of the child were paramount, the former saying that "no valid public purpose is served by stigmatizing an artificially conceived child as illegitimate,"[9] and the latter echoing this rationale.[10] Similar thinking prevails on the issue of child support—indeed, even one of the earlier cases holding AID to be adulterous nevertheless held that the husband had a duty to support the child since he had consented to artificial insemination. As far as inheritance is concerned, AID has not raised real problems here, inasmuch as the trend has been to extend inheritance rights to illegitimate children. Given the trend, supported by Constitutional decisions of the US Supreme Court,[11] it is unlikely that a court would deny an inheritance to an AID child.

What the courts are doing is, however, only one part of the story. Many states have passed statutes legitimizing AID children where both husband and wife have consented to AID. Georgia passed the first such statute in 1964. Similar statutes have been passed by Oklahoma, Arkansas, California, and New York. In 1973 the National Conference of Commissioners on Uniform State Laws produced its Uniform Parentage Act, §5 of which is as follows:

> (a) If, under the supervision of a licensed physician and with the consent of her husband, a wife is inseminated artificially with semen donated by a man not her husband, the husband is treated in law as if he were the natural father of the child thereby conceived. . . .
>
> (b) The donor of semen provided to a licensed physician for use in artificial insemination of a married woman other than the donor's wife is treated in law as if he were not the natural father of a child thereby conceived.[12]

This statute has been adopted, through 1977, by California, Colorado, Hawaii, Montana, North Dakota, Washington, and Wyoming.

As indicated earlier, SG is similar to AID in that impregnation is achieved without sexual intercourse, and in that one of the spouses is not the biological parent of the offspring. This discussion of AID seems to show that SG is not objectionable on either of these grounds alone, and in particular on the second ground alone, for there is no reason to take a different view of things merely because of the sexual reversal.

Private Adoption

Adoption is an alternative that Alice and Joe might consider—and it has the advantage of being an alternative that no state or court objects to. Indeed, it is encouraged, though it is also regulated. States regulate adoptions for the following reason: because there is no biological relation ('blood-tie') between parents and child, and because the adoptive mother has not been pregnant with and experienced the birth of the adoptive child, there is some concern about the degree to which the parents will be attached to and feel a sense of responsibility for the child. As a result, and out of a concern for the child's interests, states want assurances that the potential adoptive parents are suitable.

Some adoptions are handled by state-licensed agencies. A couple desiring to adopt a child applies to the agency, which then conducts a thorough investigation of the couple's health, financial situation, motivation for wanting to adopt, and other indicia of suitability to raise a child. When a child becomes available, the agency investigates the child and the background of its parents, trying to determine whether the adoptive couple and the child are mutually suitable.

Whatever the merits (and drawbacks—for adoption agencies have their critics) of this method of adoption, very often it is slow, sometimes taking years. As an alternative adoptions are often arranged privately. An intermediary—perhaps a friend, but often a lawyer or doctor—knows of a couple wanting to adopt, and also of a baby recently born or soon to be born that the natural parent or parents want to put up for adoption. Arrangements are made, and the child is given to the adoptive parents. In lieu of an agency investigation in these cases, states typically require some sort of investigation by state officials and appearances in court by the adoptive parents. In private adoptions there is concern not only about the interests of the child and suitability of the adoptive parents, but there are also worries about whether undue pressure has been put on the natural parents to relinquish the child. Furthermore, there are financial aspects of these transactions that need to be monitored, for there are legal fees and sometimes the natural mother's medical expenses are paid. But often a couple anxious to adopt a child is willing to pay a fee to the natural parents to give up their child, and all states are opposed to what they see as the selling of babies.

Whatever the difficulties with private adoption, they are thought to be manageable, and it is a well-established practice. Given its desirability, we can inquire as to how much it is like and unlike SG. SG has a lot in common with private adoption, for in our example Alice's sister bears the child and agrees to its adoption by Alice and Joe, who pay the legal fees and the surrogate's medical expenses. What, then, are the differences?

One important difference counts decidedly in favor of permitting SG if private adoption is permitted—namely, that in SG one of the adoptive parents is the natural father, which means that this adoption has a greater probability of success than the usual private adoption.

The most obvious difference between SG and the typical private adoption is that in the former case the pregnancy is brought about *in order to* create the adoptive situation, whereas in the latter case the child (perhaps not yet born) is already in existence. How significant is this difference? In itself it hardly seems significant. SG does increase the population, whereas the ordinary private adoption does not, but the numbers of cases of SG are never likely to be so great as to add to population concerns, and barring serious overpopulation problems, the desires of couples unable to have children in the usual way should undoubtedly have as much weight as the desires of those who can have children in the usual way. That is, if population concerns are not great enough to warrant limitations on 'ordinary' procreation, then they are not great enough to warrant limitations on SG. I do not want to be taken, though, to be saying that states have no legitimate concerns with SG—only that the mere fact that SG, but not private adoption, involves an increase in population is not an adequate reason for legitimizing one but not the other. On the other hand, the fact that the usual adoption involves children already in existence is not without significance, especially from the point of view of a state's concerns. States are undoubtedly pleased that adoption helps to fulfill the desires and even the needs of otherwise childless couples. But states are undoubtedly more concerned with the welfare and well-being of children whose outlook—a childhood spent in an orphanage—is regarded as unfortunate, and with the costs to the state in supporting such children. In adoptions in which the child is already in existence the child gains, the adoptive parents gain, and the state gains. In SG, by contrast, it is only the adoptive parents who gain, and while there is nothing wrong with this, it should caution us against being too hasty in arguing from state approval of private adoption to state approval of SG.

SG for Payment

Alice's sister is willing to bear a child for Alice and Joe without pay. Of course, they will pay her medical expenses and the legal fees, but that's all. Often, however, the surrogate is paid a fee. Many surrogates who are paid say that, though they would not do it without a fee, the economic motive is not the main reason they do it—they often say that they want to help someone, perhaps because they have known childless couples who wanted very much to have children. On the other hand, it is not unreasonable to expect that though the pioneers in SG may have other motives, others would be more interested in the money. Fees of $5000–10,000 are often talked about, and a recent television report on SG put the range at $7000–35,000.

The payment of a fee for a surrogate's time, inconvenience, and risk raises two problems that are of legal concern. The first is the matter of coercion. The character of the concern is well put in the recent case of *Doe v. Kelley*.[13] A Michigan statute makes it a felony for a person to "offer, give, or receive any money . . . in connection with any of the following: (a) the placing of a child for adoption. . . ."

John and Jane Doe, having an agreement with Mary Roe whereby the latter would, for $5000 plus expenses, carry a child for the Does and turn it over for them for adoption, sought to have the statute declared unconstitutional. The court held the statute constitutional, saying that "the right to adopt a child based upon the payment of $5000 is not a fundamental personal right."[14] But the court also went on to discuss other matters, and in the course of its opinion agreed with the point behind the query of the state's attorney: "How much money will it take for a particular mother's will to be overborne, and when does her decision turn from 'voluntary' to 'involuntary.'" "Plaintiffs," the state's attorney points out, "have initiated this lawsuit because few women would be willing to volunteer the use of their bodies for nine months if the only thing they gained was the joy of making someone else happy. . . . [T]he money . . . is intended as an inducement. . . ."[15]

If the problem is the overcoming of a person's will, there are private adoption cases where that is indeed a concern. Suppose a woman has just given birth to a child, and she and her husband, in weak financial circumstances, are worried about whether they can handle the medical bills and the impact of additional expenses on their family. An offer of money at this time might well overcome scruples they may have against giving up (abandoning, as they might see it) their child, and so perhaps there is a case for a law protecting people against themselves in a situation in which they might not be able to judge things clearly. The surrogate case is different, however. It may be true that there is an economic incentive, and that surrogates may come largely from lower income levels. But people need to be protected against themselves only when they have scruples against doing something that an offer of money might influence them to put aside. The case for overbearing someone's will cannot be made merely by showing that the individual would not have agreed but for the money—after all, that is an ingredient of all commercial transactions. What is needed is a showing that the individual thinks the activity wrong or at least questionable and is gotten to do it anyway because of the money. We may be put to the guess here, but my guess is that women who offer, or would offer, to be surrogates would not regard SG as wrong, such that their wills would have to be overborne to get them to do it.

The second problem with SG for pay has to do with whether it constitutes baby-selling, which is illegal in all states. "It is a fundamental principle," says the court in *Doe v. Kelley*, "that children should not and cannot be bought and sold." "Mercenary considerations used to create a parent–child relationship and its impact upon the family unit strikes at the very foundation of human society and is patently and necessarily injurious to the community.''[16] Issues regarding the family will be discussed later in this article; the concern right now is with baby-selling. Let us take it that baby-selling is objectionable, and that this is the main target of the Michigan statute cited earlier. And let us agree that if a couple has had a child, and takes money to consent to its adoption, that is a case of baby-selling. Does SG similarly fall under this heading?

Whatever doubts people may have about SG, it is hardly likely that anyone's real concern is that it amounts to selling babies. After all, the 'purchaser' is (or at least one of them is), in the cases we are considering, the natural parent of the child, so that there are already the ties of affection and concern that go along with parenthood. To regard this as an outright purchase is to regard the father as buying his own child, which is a strange way of looking at the matter. Furthermore, in these cases there is only one possible 'purchaser,' which is hardly the model of a commercial transaction. As far as I can see, this is not a purchase of an object, or commodity—namely, the baby—at all; it is the purchase of a service—the service of conceiving, carrying, and giving birth to a child for another couple.

The words of the Michigan statute do, however, apply to the arrangement. Did the Michigan legislature intend to bar SG for money? Probably not, inasmuch as it is unlikely that anyone thought of SG when the statute was passed. Should a court apply a statute to a situation not contemplated by the legislature? Statutory interpretation is of course not a science, and there is room for judicial maneuver in such cases. But it is usual for a court, in seeking the intent of the legislature, to read and apply statutes literally and not try to guess what the legislature would have done had it considered the issue. This strategy further recommends itself to a court when the outcome is on the conservative side, particularly when an issue is controversial: better to leave it up to the legislature to break new ground.

It is important to note that one of the main reasons the baby-selling issue comes up at all is that the law regards the natural mother (to be understood here as a woman who is both genetically and gestationally linked to the child) as parent, but not the natural father. And if the surrogate is married, her husband is presumed, though rebuttably, to be the natural father. The result is that the natural father must adopt his child, and, as the (sketchy) law in this area is presently interpreted, has no legal claims with respect to the child at all. Indeed, it is an irony that the language of §5(b) of the Uniform Parentage Act (quoted above), in trying to deal with some of the problems of artificial insemination by donor—specifically, to prevent sperm donors from having to bear legal responsibility for children produced by the use of their sperm—creates additional problems for SG, for the natural father of a child born of a surrogate is a semen donor within that statute, and thus is to be "treated in law as if he were not the natural father of a child thereby conceived." Here is another case of a statute designed for one purpose applying to a situation not contemplated when it was drafted and passed.

At this point we are in the neighborhood of some of the practical legal problems standing in the way of SG and needing to be worked out if the practice is to be recognized.

Other Problems

There are a number of legal problems that do not raise difficulties of principle, but which would call for the revision of several laws in order to accommodate SG—if, that is, it comes to be regarded as acceptable. Section 5(b) of the Uniform Parentage Act, referred to above, is a good example of the sort of redrafting that would be needed. No opinion is expressed here as to whether such tasks are exclusively for legislative action, or whether instead there are some matters with respect to which courts should step in. Certainly, however, there is little that a legislature could not do, if it chose to pave the way for SG.

Consider, for example, the following items of Kentucky law,[17] which, according to the Attorney General of Kentucky, render SG contracts illegal and unenforceable in that state, though SG was not contemplated when they were passed. (1) KRS 199.500(5) says "In no case shall an adoption be granted or a consent for adoption be held valid if such consent for adoption is given prior to the fifth day after the birth of the child." (2) KRS 199.601(2) says: "No petition may be filed [for voluntary termination of parental rights] prior to five (5) days after the birth of a child." (3) KRS 199.590(2) says: "No person, agency, or institution not licensed by the department may charge a fee or accept remuneration for the procurement of any child for adoption purposes." (4) KRS 199.590(1) says "No person . . . shall advertise . . . that it will receive children for the purpose of adoption nor shall any newspaper . . . contain an advertisement which . . . solicits the custody of children." This provision would, I assume, make it illegal in Kentucky for a person to place, and for a newspaper to carry, the following advertisement, which appeared in a student newspaper in California: "Wife is looking for a healthy, blue-eyed woman who is willing to carry my husband's child. All expenses paid plus $4000."[18] There are in addition other laws, in Kentucky and elsewhere, which, as drafted, seem to stand in the way of SG arrangements.

It is fairly clear what steps would have to be taken to revise these laws if SG were to be permitted. Less clear, and of considerable interest, is the matter of breach of contract (where SG rests on an enforceable agreement): what happens if the surrogate refuses to turn over the child or wants to terminate the pregnancy, or the contract parents refuse to accept the child?

There are two sorts of cases of breach by the surrogate to consider. First, the surrogate might abort, not for any medical reason but merely because she no longer wishes to go through with the agreement. It is an interesting and important question whether the contract parents would be able to enjoin the abortion—some lawyers have expressed the view that no court is likely to enforce a contractual provision requiring a woman to refrain from a desired abortion.[19] But that does not mean that a surrogate who aborts has no liability to the contract parents for damages caused thereby, though there is still the further question of what the elements of damages will be. The second sort of case of breach by the surrogate occurs when she wants to keep the child herself and refuses to turn it over to the contract parents. Damages are again a possible remedy, but the real question is whether the surrogate can be required to perform her end of the bargain. Elements of such disputes are analogous to custody disputes, which are handled by determining where the best interests of the child lie. There has been, in such cases, a feeling that a child is best left with its natural mother, though support for this view has been eroding lately. If indeed there is no real basis for favoring the natural mother, and if both families (the contract parents, and the surrogate and her husband) seem equally suitable, a plausible suggestion is that the terms of the contract should prevail, in that it is an agreement entered into voluntarily by the parties.[20]

Breach of the contract by the contract parents is in some ways a more serious matter, for the result might be that neither parent wants the child, which is of course a very bad thing for the child and a concern of the state, which could wind up having to support the child. But there are two points that mitigate this concern. First, breach by the contract parents is an unlikely event—at least where the child is not abnormal—inasmuch as they have sought a child for a long time and have resorted to surrogation only when all else failed, and inasmuch as the child is the natural offspring of the contract father. Second, the child can be put up for adoption, and if unadoptable the contract father can be held liable for its support.

There are a number of other questions that can arise between contract parents and surrogate, most of which can be handled by having a carefully designed contract. Nevertheless, there are as yet untested questions. May the contract restrict the surrogate's diet, drug intake, activities, doctor, or number of visits to the doctor? May the contract require intrusive procedures such as amniocentesis or other methods of fetal monitoring? May the contract parents require the surrogate to have an abortion if the fetus is discovered to have a defect? Some lawyers venture the opinion that courts would not enforce provisions of a contract requiring abortion or intrusive procedures.[21]

Morality

Not much has been written on the morality of SG. Many writers, to be sure, *say* that there are important moral questions. But those questions that are usually raised are more tangential ("What would happen if. . . .?") than central. For example, it has been asked whether the contract parents and the surrogate should meet, whether single women should be surrogates, what degree of control the contract parents can exercise over the surrogate, and whether a single man may contract with a surrogate. These are genuine questions, of course, and some of them may be important. But the real question of the morality of SG can be addressed only by focusing on the most favored cases—those in which none of the tangential questions are at issue.

The Case for SG

The case for SG is quite simple: it is that everyone benefits from the arrangement. The benefit to the contract parents is obvious. Presumably they have tried for a long time to have a child and have tried whatever offered hope—therapy, fertility drugs, or whatever. Finally they have been willing to go to the trouble and expense of SG as the second or even third best solution to their problem. In almost every way that people regard the having of children as a benefit, contract parents expect to benefit in just the same ways.

The surrogate may benefit in any of a number of ways. There is, of course, an economic benefit if a fee is involved. If there is no fee, the surrogate undoubtedly gains the satisfaction of helping someone; indeed, this satisfaction may be present even if there is a fee—many people in service, and even in sales, professions seem to gain such satisfactions along with their pay. Again, if no fee is involved, a surrogate may gain the satisfaction of performing a supererogatory act. Finally, and very important for many surrogates, many women regard pregnancy as a desirable experience and the bearing of children as a creative act—and for some this is apparently the case even if they are not going to keep the offspring they bear.

Some writers even claim that not only do the contract parents and the surrogate benefit, but the child thereby produced also benefits. The argument has been put this way: ". . . one might argue that the child would never have existed had it not been for the surrogate arrangement, and so whatever existence the child has is better than nothing."[22] Again: "Without the new modes of technology the aritificially conceived child would have no existence at all. Surely it is better for him to exist than not."[23] Is this a sound line of thought? In one sense it certainly is better to exist than not; even though it occasionally happens that a person prefers to die rather than remain alive, most people prefer life. But it seems strange to say that it would be better for a nonexistent entity if it were to come to exist. This strangeness is what is behind the Yiddish joke reported by Robert Nozick in *Anarchy, State, and Utopia*: "Life is so terrible; it would be better never to have been conceived."—"Yes, but who is so fortunate? Not one in a thousand."[24] The problem is that to say it would be better for the SG child to exist rather than not is to presuppose an entity of whom we are speaking. But prior to conception there is no entity in existence of whom we can make that statement. The picture that is conjured up is that of a shadowy being in some sort of metaphysical limbo, who will come to exist in physical reality if, but only if, a surrogate will bear it, and who will thereby be better off. This picture is obviously in error.

Even without the last supposed benefit of SG, however, the benefits it offers the contract parents and the surrogate make a powerful case for the practice.

The Case Against SG

I have been able to identify four arguments against SG, of which I regard the third as the most important.

Unity of Love and Procreation

This is an argument deployed by conservative religious moralists, specifically in the Christian tradition. The claim is that "acts of sexual love should never be non-procreative."[25] This is not interpreted to mean that one must procreate, nor that one must want to procreate or desire that particular acts result in pregnancy. Nor does it even require that one be fertile: "sterile individuals can behave procreatively."[26]

Apparently all that it means is that people must not deliberately deprive their sexual intercourse of its 'procreative character.' The reason for this has to do with the 'meanings' of sexual acts. Sexual intercourse has a unitive meaning and a procreative meaning, and, the claim is, these meanings are inseparably connected. Thus, a sexual act deprived of its procreative character is illegitimate, and this rules out artificial contraception, artificial insemination using donor sperm, and, of course, SG. But why, one might ask, is it not good enough for these sexual and procreative meanings to be separate elements of a single, exclusive relationship, rather than present in each and every act done by parties to a relationship? A contemporary religious moralist, we are told,[27] argues that just as in the creation God combined a loving and a creative act, so in a marriage one must not have two different relationships—one loving and the other procreative—but one relationship that is both loving and procreative at the same time.

Without attacking this entire tradition, and interpreting the last bit as not ruling out contraception, I would register two doubts about the line of thought sketched above. First, even within the confines of this tradition it seems to me that what is being described is an ideal, something to which people should perhaps aspire. But having identified an ideal, the argument proceeds to say that what is less than ideal is wrong. As a general matter, however, that which is less than the best is not, for that reason alone, wrong, and so a stronger argument is needed for declaring wrong the separation of love and procreation in the most favored instances of SG.

Professor David Smith, on whose article I have relied for the argument being discussed, illustrates the demand of the unity of love and procreation as follows: "This principle would be violated in the case of a man who maintained a marriage for breeding and reasons of status while keeping a mistress on the side."[28] In this case, though, one has a choice—one can, if he wishes, localize his love and procreation in a single relationship. Not so where SG is an issue. Well, then, even if we accept the idea that love and procreation should go together, and even if we regard this as a basis for rejecting the behavior of the man described above who has one relationship for love and another for procreation, must we accept that it applies to SG as well? Why should we not regard impossibility as mediating the application of the principle of the unity of love and procreation? What I mean is this: We can accept as a general principle the idea that love and procreation should go together, but limited to those cases where it is possible for them to exist together. There is no compelling reason to insist on it in cases in which it cannot be realized.

The second doubt I would raise is that it seems to me to be unremittingly perverse to say that if a couple is unable to have all three desirable things (i.e., love, plus child, plus unity of love and procreation) because it is unable to have the third of them, then it cannot have both of the other two desirable things, but must make do with only one of them.

Analogy to Prostitution

It is argued that ''allowing one's womb to be used by another couple is analogous to allowing use of one's body solely for the sexual pleasure of another.''[29] And when the use is for pay, the situation is seen as similar to prostitution: ''. . . to bear another's child for pay is in some sense a degradation of oneself—in the same sense that prostitution is a degradation *primarily* because it entails the loveless surrender of one's body to serve another's lust, and *only derivatively* because the woman is paid. It is to deny the meaning and worth of one's body, to treat it as a mere incubator, divested of its meaning.''[30]

Whether SG denies the meaning and worth of one's body is a matter that is very much in the eye of the beholder. Apparently most of the surrogate baby-bearers to date do not feel this way, or feel in any other way degraded. Most of them seem to feel that the experience fulfills them, and that using their bodies so as to benefit others in this way is not only all right, but positively heroic.

There can be no objection to using one's body for another's ends as such, for, after all, many people labor with their hands for pay, which is as much a use of one's body as any other. Undoubtedly proponents of the prostitution analogy are thinking only of certain parts of one's body—the parts connected with sex. It is using these parts for pay that is regarded as degrading. Let us grant that prostitution is degrading and objectionable for the reasons given above. That is, let us grant that using one's genitals for pay in a loveless surrender to serve another's lust is degrading. How does it become analogously degrading to use a different part of one's body (the uterus) for entirely different purposes? The only connection between genitals and uterus is that they are parts of the body and are connected with reproduction. A uterus has nothing to do with lust.

Roots

For many of us the matter of who we are is bound up with where we came from—that is, with our ancestry. There has been a concern with roots in recent years. Adopted individuals in particular have sought in-

formation about who their biological parents are, and often characterize their quest as a search for self-identity.

In the usual adoption case, where the child is already in existence,
there may be no way to avoid these dislocations, though many writers
maintain that adoption practices should be modified so as to permit
contact with the biological family.[31] But with SG a choice is being
made—one *decides* to bring into existence an individual who will suffer from a deficiency, namely, being estranged from one of its biological parents, and thus having its identity in part compromised. The case
against this is put most strongly by Leon Kass:

> Properly understood, the largely universal taboos against incest,
> and also the prohibition against adultery, suggest that clarity
> about who your parents are, clarity in the lines of generation, clar-
> ity about who is whose, are the indispensable foundations of a
> sound family life, itself the sound foundation of civilized commu-
> nity. Clarity about your origins is crucial for self-identity, itself
> important for self-respect. It would be, in my view, deplorable
> public policy further to erode such fundamental beliefs, values,
> institutions, and practices. This means, concretely, no encourage-
> ment of embryo adoption or especially of surrogate pregnancy.[32]

There are two lines of thought to pursue with respect to this argument.
First, how strong is the evidence that lack of information about one's
biological parentage creates problems for individuals? And how great
are these problems? The place to look, of course, is to the situation of
adopted children. There are two sorts of data to look for: first, evidence
of psychological disturbance, difficulties in adjustment, problems in
personal relationships, and so on, and second, the search for roots, the
attempts of adoptees to find out about their backgrounds.

As to the first of these, there is a fair amount of evidence of such
problems among adoptees. But though the claim is made that it is the
very fact of being separated from one's lineage that is responsible for
these problems, the cases that are usually offered as evidence involve
individuals who lived for a period of time—often lengthy—with their
biological parents, and then were separated from them. Such individuals often experience psychological problems involving a lack of a sense
of emotional support, loyalty, and the like, but given the character of
the evidence it is difficult to associate this wholly (if at all) with lack of
biological tie. It is clear also that a lot of the difficulty that adoptees
have derives from fears (genuine or imagined) about how others, such
as friends or fiancés, will react to their adoptive status.

It is true, certainly, that some adoptees do have a deep desire to learn about their biological backgrounds. It is unclear the extent to which this leads to other psychological problems for these individuals, but let us grant that it does sometimes produce a diminished sense of identity. How great a problem is this? It is hard to say, though the efforts that some go to to learn who their biological parents are is important testimony on the matter. This leads to the second line of thought. Assuming that the SG arrangement will result in a child that may suffer from a partial lack of a sense of identity, is this possibility so serious that the contract parents should not seek to have the child at all? I do not think that this view can be maintained. One could as well argue that people ought to refrain from having children that will be born poor (though not wretched) or members of disadvantaged minorities. There are, of course, circumstances in which people should not have children, though the cases are not at all clear. Where parents are likely to pass on a serious hereditary defect, they should not have children. Or if it is known that in other ways a child will have no chance at all in life, one should not have a child. But suppose we cannot afford the best schooling, or that the child will be born with a mild physical handicap, or that one parent has personality traits that are likely to lead to his or her child being ill-tempered. It seems to me that the potential for problems with one's sense of identity falls into the latter group of cases— things to be concerned about and to try to overcome, but not serious enough to make it wrong to have a child by SG.

Family

Some writers believe that SG and related arrangements are detrimental to the family. A clear statement of this objection, referred to above, comes from Leon Kass: ". . . clarity about who your parents are, clarity in the lines of generation, clarity about who is whose, are the indispensable foundations of a sound family life. . . ."[32] And since, as Kass sees it, sound family life is an essential of "civilized community," SG threatens civilization itself. Though in itself this is a bit much to swallow, I take Kass to be saying that *too* little clarity about blood lines is detrimental to organized social life. Though this may be true, the means of assessing it are beyond me, and I will assume, for the sake of argument, that SG and related arrangements will never become a problem of that magnitude.

In what ways, then, might SG have adverse impact on the family as a social institution? Unfortunately, those who are protective of the family seldom make clear what the problems are, so we must speculate for ourselves. One potential source of difficulty is the fact that the SG child is adopted by one of the parents. But this would most likely become significant only when other elements of the family relationship had begun to deteriorate. Nevertheless, it is always a potential sore spot.

When we think of the role that we expect families to play in social life, the picture we get, it seems to me, is that there is little to fear from SG. One important job of the family is to provide an appropriate setting for developing children's social skills and raising them to responsible adulthood, a job which is in no way compromised by SG. Another function of the family is to provide mutual support, especially love and emotional support, for its members; this too is in no way compromised by SG. Families have also traditionally done the job of caring for the sick and the old. Undoubtedly the family has been losing ground to the state in these areas, and probably some of the same forces and trends that have led to the possibility of SG have been responsbile for this diminished role of the family. But having common origins does not make SG (or anything else) responsible for this situation, and prohibiting SG would in no way stem the tide. Indeed, in terms of this role of the family, SG should probably be encouraged, in that it offers the possibility of familial rather than state support for an otherwise childless couple as they age. The following, from Justice Traynor of the California Supreme Court, is an eloquent statement of the role of the family:

> The family is the basic unit of our society, the center of the personal affections that ennoble and enrich human life. It channels biological drives that might otherwise become socially destructive; it ensures the care and education of children in a stable environment; it establishes continuity from one generation to another; it nurtures and develops the individual initiative that distinguishes a free people.[34]

None of this is compromised by SG (depending, that is, on what sort of continuity Traynor is talking about).

One element of the family that SG cannot preserve, however, is purity of blood lines (at least, it cannot be preserved on both sides). I am not in a position to speculate on all of the reasons why some people regard this as important—probably such matters as status, social position ('background'), a sense of history, and maybe esthetic responses are part of the picture. The only observation I would offer is that it seems pointless to worry about impurity of blood lines when the alternative is for that particular branch of a family tree to end altogether.

Conclusion

As far as I am able to determine, there is not much of a legal case or moral case against SG. There are legal problems, to be sure, but these can be worked out if the public wishes to permit SG. It is likely, though, that people will resist SG. It is very different. But it is hard to put one's finger on anything that is wrong with it. Probably there is a strong element of the idea that it is not 'natural,' and that what is not natural is wrong or at least suspect, that is behind people's doubts about it. The notion of the natural has been discussed in many other places, and there is no need to take it up here. For myself, I cannot see that the unnaturalness of SG amounts to anything more than its being unusual, and there doesn't seem to be any moral difficulty in that.

Notes and References

[1]381 U.S. 479 (1965).

[2]316 U.S. 535 (1942).

[3]405 U.S. 438, at 453 (1972).

[4]431 U.S. 678, at 685 (1977).

[5]Hoch v. Hoch, No. 44-C-8307, Cir. Ct. Cook County, Ill. (1945). Cited in Mary Ann B. Oakley, "Test Tube Babies: Proposals for Legislative Regulation of New Methods of Human Conception and Prenatal Development," in Sanford N. Katz and Monroe L. Inker (eds.), *Fathers, Husbands and Lovers: Legal Rights and Responsibilities*, Publication of ABA Section of Family Law, 1979, p. 77.

[6]Doornbos v. Doornbos, 22 U.S.L.W. 2308 [Super. Ct., Cook County, IL (1954)]. Cited in Oakley, *op. cit.*, p. 77.

[7]MacLennan v. MacLennan, Sess. Cas. 105, (1958) Scots L.T.R. 12. Cited in Oakley, *op. cit.*, p. 77.

[8] People v. Sorensen, 68 Cal.2nd 285, 437 P.2d 495 (1968).

[9]437 P. 2d, at 501.

[10]In re Adoption of Anonymous, 345 N.Y.S.2d 430 (Surrogate's Court, 1973).

[11]See Weber v. Aetna Casualty & Surety Co., 92 S. Ct. 1400 (1972), and Gomez v. Perez, 93 S. Ct. 872 (1973).

[12]*Uniform Laws Annotated: Matrimonial, Family and Health Laws* (St. Paul, West Publishing Co., 1979), pp. 592–3.

[13]6 FLR (Family Law Reporter) 3011.

[14]6 FLR, at 3013.

[15]6 FLR, at 3014.

[16]6 FLR, at 3013.

[17]7 FLR, 2246–7.

[18]*Daily Californian*, Nov. 19, 1976, p. 19. Quoted in Elizabeth A. Erickson, "Contracts to Bear a Child," *California Law Review* 66 (1978), 611.

[19] See Dennis M. Flannery et al., "Test Tube Babies: Legal Issues Raised by *In Vitro* Fertilization," *Georgetown Law Journal* 67 (1979), 1317.

[20]See Erickson, *op. cit.*, p. 621.

[21]Flannery et al., *op. cit.*, p. 1317.

[22]George J. Annas, "Contracts to Bear a Child: Compassion or Commercialism?" *Hastings Center Report* 11 (1981), p. 24.

[23]David H. Smith, "Theological Reflections on the New Biology," *Indiana Law Journal* 48 (1973), pp. 605–22. It is not clear whether Smith accepts this argument.

[24]Robert Nozick, *Anarchy, State, and Utopia* (New York: Basic Books, 1974), p. 337.

[25]Smith, *op. cit.*, p. 619.

[26]*Ibid.*

[27]Smith, *op. cit.*, p. 621.

[28]*Ibid.*

[29]*Ibid.*

[30]Leon R. Kass, " 'Making babies' revisited," *The Public Interest*, No. 54 (Winter, 1979), 47.

[31]See Fernando Colón, "Family Ties and Child Placement," *Family Process* 17 (1978), 289–313.

[32]Kass, *op. cit.*, p. 47. Kass is, however, very cautious in his recommendations. He says it would be foolish to try to outlaw SG, but that government should not "foster" such practices.

[33]Kass, *op. cit.*, p. 47.

[34]DeBurgh v. DeBurgh, 39 Cal.2d 858, 250 P.2d 598, at 601 (1952). Quoted in Flannery et al., *op. cit.*, p. 1315.

Surrogate Motherhood
The Ethical Implications

Lisa H. Newton

God, Motherhood, and Promethean Terror

When we initially heard about the first test-tube baby, two reactions surfaced immediately: first, congratulations and happiness; second, condemnation and alarm. Why? Because as we have always known, motherhood is sacred. It is so valuable that any means that we can develop to bring it to those for whom it was formerly impossible are blessed means, cause for the most sincere and total joy; and it is so holy and mysterious that any voluntary human intervention in the process is cause for terror of divine wrath. The most primitive and powerful hopes and fears of the human race hover around all reproductive technologies, and all new social or legal arrangements made possible by them, "Surrogate motherhood" included.

To engage the ethical issues involved in surrogate motherhood, then, is to take on the Western ethical heritage at the point where it disappears into its tribal past, becoming indistinguishable first from religion, then from superstition. The first purely ritual objects used by humans to invoke divine powers were, to the best of our knowledge, devices of Motherhood: squat female figures, hugely pregnant, grotesquely shrunken in head and limbs, pure gestation. "Motherhood" has never had a straight, clear, and neutral meaning, devoid of connotations, from which we might begin to piece together its moral history—the power to bring forth a totally new human being, apparently *ex nihilo,* has always been an object of superstitious awe. The topic of Motherhood, viewed as a topic, is completely ungovernable; if

I am to say anything at all sensible about the ethics of its surrogate (literally, "that which is put forward as an alternative" to it), I am going to have to restrict the subject matter and the questions I will ask about it.

My inquiry is sadly limited by the lack of available literature on the subject. The last serious treatment of the topic, as far as I am able to discover at this writing, is the first chapter of Angela Holder's *Legal Issues in Pediatrics and Adolescent Medicine*, published before the first "test-tube baby" was successfully implanted by Drs. Steptoe and Edwards. A recent casebook, Shapiro and Spece's *Bioethics and Law*, contains excerpts from the one reported case on surrogate motherhood, *Doe v. Kelley*, plus other materials of a popular nature. Beyond that, we are restricted to a few *New York Times* Magazine section articles and some other minor pieces in the popular press.[1]

Very well then, as far as subject matter is concerned, excluded from the topic of this paper are the following practices:

1. *Adoption*. Adoption, especially by godparents after the death of natural parents, must be the oldest form of surrogate motherhood known to the race. All provisions for the care of the born young in the absence or incapacity, temporary or permanent, of the biological mother, will be similarly excluded—wet nursing, babysitting, day care, provision of state-run orphanages, and all their relatives. It will be assumed that these practices are all perfectly acceptable as long as the child is not neglected or exploited, and as long as the actual care giver has, *as at least one motive* in undertaking the care of the child, concern for the welfare of the child. As the argument progresses I shall have to ask whether this concern must be the *sole* motive in undertaking such care. (When we pay the babysitter, what are we paying her *for*?)

2. *Artificial Insemination by Donor* (AID) and in vitro fertilization with implantation of embryo, at the eight-to-sixteen cell stage, into the woman who is at once donor of the egg and prospective guardian of the child. In vitro fertilization as described (the procedure accomplished successfully by Drs. Steptoe and Edwards) I take to be merely an elaborate way of routing the mother's ovum through the natural events of ovulation, fertilization, and implantation.[2] AID modifies fatherhood, not motherhood. In both cases, the presence of twentieth-century technologies may raise ethical problems, but not problems of surrogate motherhood, since in these cases the mother stands for herself, without substitute. Similarly all uses of drugs or mechanical devices to aid pregnancy, or a certain type of pregnancy, in the woman who will conceive, bear, and raise the child, are excluded from the purview of this paper.

The procedures in dispute, then, involve only embryos, fetuses, children, between conception and birth; the surrogate mother stands in for the mother not in her roles of nurse, care giver, and guardian, but in her roles of ovulator and gestator—of generator of the egg and of carrier of the child in the womb. These, of course, were the very roles in which woman has been an object of worship and ritual fear. Two different surrogations are in question here, with different consequences. The simpler surrogation is that of ovulator: a fertile woman donates a normally ovulated egg obtained from her body by some unspecified procedure, or a woman whose ovary must be surgically removed donates an egg obtained during removal, to a woman who is unable to ovulate normally; this egg is fertilized in vitro, possibly with the sperm of the infertile woman's husband, and the fertilized zygote implanted in the uterus of the infertile woman, who carries it to term, gives birth to the infant, and assumes the roles of nurse and guardian. The more complex surrogation involves surrogate gestation: the ovum, which may be obtained from the prospective guardian of the child, from the prospective carrier of the fetus, or from an unrelated source, is fertilized, presumably with the sperm of the prospective guardian's husband, implanted in the uterus of the carrier, who carries it to term, gives birth to the infant, and then gives the child up to the nurse and guardian. Where the egg donor is the prospective guardian or unrelated to either guardian or carrier, the fertilization takes place in vitro; where the egg donor is the prospective carrier of the fetus, the fertilization can be carried out in vivo, by standard artificial insemination procedures, with a much higher probability of success. In this case, only the contractual arrangement concluded prior to pregnancy, specifying that that baby is to be surrendered by the carrier to the prospective guardian at birth, possibly for a fee, differentiates the situation from the more familiar one of giving a newborn child up for adoption.

These two situations, then, surrogate ovulation (egg donation) and surrogate gestation (host-motherhood), particularly the latter, will occupy our attention in this paper. We wish to discover the ethical problems inherent in these situations, the *loci*, as it were, of our intuitive moral doubts concerning stand-in motherhood.

If this is our subject matter, what are our concerns? We are dealing with formless doubts and terrors, but also with mundane considerations of social policy and expense.

Given the presumption that normal motherhood is to be viewed as an unequivocal good (which, of course, is open to question on ecological grounds beyond the purview of this paper), the ethical doubts surrounding laboratory-based surrogate motherhood may, I think, be summarized as follows; it is feared that:

1. Some moral essence of motherhood is lost or terribly endangered in the laboratory procedures, or
2. Some immoral act is (necessarily) performed in the course of
   the laboratory procedures, or
3. In the case of surrogate gestation, morally intolerable exploitation of the surrogate or the child is (necessarily) involved; or
4. Social resources are unjustly allocated in the expenditure of
   money and energy to pursue the techniques of surrogate motherhood, or
5. Social resources are inefficiently expended in this technology,
   or
6. Any combination of the above.

The order of presentation of items 1–5 is an order of decreasing moral
complexity. Item 5 is susceptible to some approximation of simple
cost-benefit analysis; items 2, 3, and 4 are susceptible to measure on
reasonably accessible criteria; item 1 will take a more extended analysis just to phrase the question sensibly.

The purpose of this paper is to explore the ethical concerns surrounding laboratory-assisted surrogate motherhood; the method of the
exploration will be to examine how those concerns apply to surrogate
motherhood, as described. Before undertaking the exploration, we can
focus our problem by taking note of a form of surrogate motherhood
that we will *not* be considering. After all, substitution in such matters
need not take place in the laboratory. If a barren woman wishes a child
by her husband, and cannot conceive and/or bear it herself, and finds
some healthy young woman willing to do the conceiving and gestating
for her, why not simply send her husband over to the woman's bed at
the appropriate time of the month to fertilize the ovum in the ordinary
way? I am not sure whether the good citizens of the United States are
prepared to accept such a practice as a legitimate way of obtaining babies; I am reasonably certain that some of the concerns that arise in
laboratory-assisted surrogation, for instance over presumptive custody
of the child and the legitimacy of payment (see below) would arise in
this case also. But I am absolutely certain that the major qualms about
this practice would be different from the fears encountered where laboratory equipment plays a necessary role in maternity. The major problem in the case supposed is one of out-and-out adultery, albeit adultery
acquiesced in by the party most likely to complain of injury from the
deed, not a brand of evil entirely new to us[3]; but the problems that
bring surrogate motherhood into contemporary discussions of bioethics
are precisely those entailed by the most recent developments in the
most sophisticated of biotechnologies. The good old fashioned forms

of vice, and their consequences, we can cope with, if not always very sensitively[4]; the newer forms present terrible problems precisely because the vice is obscure. In an embryo transfer following in vitro fertilization, or AID, there is no obvious adultery—no prurient interests, no sinful act, no stolen pleasure (no pleasure at all, in fact)—just a scientific breakthrough, test tubes and white coats in a sterile hospital, and the numbing impersonality of medical technique. If we could find someone having fun, would we know whom to blame? But all we find is a "disease," infertility, physicians specialized in "treating," possibly "healing" the disease, the instruments and substances appropriate to the medical task, and a "procedure"—the result of which is pregnancy that is not a product of normal marital relations. There is a lingering uneasy sense that somewhere, somehow, someone was taken in adultery (although the courts have laid to rest any literal fear to that effect, for normal AID pregnancies),[5] and that the whole affair has been rendered invisible and removed from vulnerability to condemnation by enclosure in the stainless world of medical science. All of this is very different from the adultery, adultery with the wife's consent and for good purpose, that might seem an obvious solution to the problem of infertility. The laboratory transforms the moral contours of the situation, to the point where sexual reproduction, motherhood, the most common phenomenon of the human experience, is suddenly difficult to grasp. In pursuing techniques to provide children for the childless, we may, simply and commendably, be carrying on the task bequeathed to us from Aesculapius and Hippocrates, to use our God-given intelligence to improve the human condition. Or on the contrary, we may, in these attempts to impose human whims and wants on the most basic of life processes, be committing an intolerable trespass on domains best left to the working of God and nature; we may stand condemned with Prometheus for making off with the property of the Deity. In the section that follows we will attempt to work out the implications of the new motherhood for the moral concerns set forth above, to see whether we stand more appropriately with Aesculapius or with Prometheus in the eyes of Zeus.

Essays on Justice and Utility

We do not expect to solve the problems of surrogate motherhood in this paper. But subjecting surrogation, and its attendant physical and social contexts, to the questions raised above, should at least give us a clearer view of the relative seriousness of the problems and the morally relevant factors that will have to be dealt with in any future solution. Unless otherwise or additionally specified, the form of surrogate motherhood under consideration will be that in which an egg is taken from the prospective guardian (donor-mother) of the child, fertilized (presumably with the sperm of the prospective guardian's husband), and implanted in the womb of the gestator (host-mother), who by arrangement (presumably contractual, presumably for pay), carries the child to term and turns it over to the guardian who then raises it as her own child. The moral dimensions on which the practice is to be measured are the five mentioned: presence of some essential violation of motherhood; necessity of acts contrary to the moral law; exploitation of gestator and child; justice; and efficiency. We shall take these, for simplicity's sake, in reverse order.

The question raised under the heading of efficiency is really a very simple one: given that that woman, or couple, who will assume guardianship of the child, initiate the surrogation negotiations because they want a baby, is there not a more efficient (cheaper, easier) way to produce a satisfactory baby for them? Why, after all, do they not simply use the procedures already available to adopt a baby already born? The answer seems to be that there are not enough healthy babies currently available for adoption through public agencies, and/or that the ones that are available, usually through private sources, are too expensive to obtain.[6] Meanwhile the abortion rate is soaring. The answer is obvious: use some of the money now going to fertility and embryo-transplant research to induce already pregnant women (presently planning abortion) to carry their babies to term. Since the women clearly do not want the babies, they should certainly be willing to give them up for adoption at birth, and the supply of babies available for infertile couples would be increased. The money for the inducements, drawn from funds (mostly governmental) now going into fertility research, could be funneled into the enterprise through a variety of agencies, primarily the hospitals where abortions are performed. It could be made available to the pregnant women by the counseling services of the abortion clinic, targeted especially to second-trimester aborters (for whom the abortion is more difficult, who have already suffered some of the penalties and discomforts of pregnancy, and who have a shorter time to term), and directed to a variety of uses: compensation for lost employment, living and medical expenses of pregnancy and childbirth, or simple cash payment for enduring rather than ending the pregnancy. Problems of justice arise immediately, of course: why should not women conscientiously opposed to abortion, who presently carry babies to term and give them up for adoption, share in the largesse? And will payment be the same for all, based on need, or adjusted to the value the pregnant woman places on her time and convenience—i.e., to the price she demands? But our present objective is to examine the possibility of raising the number of babies available for adoption, not to enact perfect justice. And paying pregnant women to carry to term, rather than to abort their babies, should be a step in the right direction.

The solution is obvious, but not, perhaps, ideal. There are, for one matter, difficulties of implementation beyond the questions of justice. Not all women can carry pregnancies to term; where the woman's or the baby's health is problematic, probably the pregnancy should not go to term merely for the sake of providing a baby for the adoption market; there may well be unforeseen psychological difficulties in renouncing a baby once born, even a baby that was at one point to have been aborted. For another matter, prospective guardians may be determined to have *their* child (biologically), and may refuse to be reassured by the observation that adopted children generally come to resemble, even physically, their adoptive parents as much as other children physically resemble their natural ones. Should the parents not have the right to exercise that preference?

Possibly they should not, which brings us to our second category of moral qualms about surrogate motherhood. Given worldwide problems of overpopulation, given the poverty and sickness of human beings already conceived and, to their misfortune, born, given the present scarcity of dwindling resources, how can research into embryo transplant technology be defended? Again, public funds may be heavily involved in support of the research; where it is not being carried out directly on government grants, at least the institutional homes for the research enjoy tax exempt status. Ought the public to be paying for the gratification of selfish whims by a few infertile, wealthy, and slightly neurotic couples who wish to have "their own" babies? Although there is surely no harm in allowing those with the money to do so to purchase any medical technology that is readily available for any reason they like (e.g., cosmetic surgery purchased solely for reasons of vanity), it is quite another matter to allocate resources to its development. Infertility, after all, is not a life threatening condition.

If injustice is a clear concern in the relevant macroallocation issues, exploitation is also a concern in microallocation—specifically, in the treatment of the gestator and (occasionally) the baby.[7] (For some reason, the literature reflects no concern at all that the prospective guardians, the couple desperate for their baby, may be undergoing exploitation by the technology publicists.) The period of concern here is the gestation of the fetus through birth and surrender to the guardian;

the laboratory techniques themselves that led to the pregnancy are not in question. The problem that arises can be seen in one of two ways, depending on your view of the humanity (''personhood,'' perhaps) of the embryo: If the embryo at the eight-to-sixteen cell stage is not a person, nor will be fully one until birth, then only certain genetic materials are implanted into the gestator. She grows a baby, gives birth to it, and it is hers. If she then hands it over to another woman, and accepts a sum of money for the transfer, then she is selling the baby. The baby is thereby subjected to the greatest exploitation known to man, viz., slavery—it is being treated as an object eligible for purchase and sale. This is flatly illegal in evey jurisdiction,[8] and no case of surrogate motherhood would ever be described in this way. If on the other hand, the embryo is (ontologically) a full human being at implantation, or becomes one between implantation and birth, the situation is quite different. In this case, the prospective guardian adopts the embryo prior to implantation (the pure case) or sometime during pregnancy (the mixed case). The gestator merely carries the guardian's baby, delivers it, and it is the guardian's. The gestator, the ''surrogate mother'' is then paid for the service she has rendered in caring for the adopted child while the real mother, egg donor and guardian, was incapable of caring for it herself. I take it that this is the core case of ''surrogate motherhood'' assumed by the popular press in its occasional exuberant explorations of the subject. What abuse, real or potential, makes this practice so controversial?

In law, the whole practice is suspect, and may be in fact impossible at present, simply because of the law's tradition of not treating a child as a full human being, with fully actualized rights, until it is born. For this reason judges trained in the law are not likely to give serious attention to arrangements that treat the fetus as an identifiable person for whom childcare arrangements could be made, and are invariably skeptical of prebirth adoption arrangements. They are very likely to assume that the gestator, delivering the child to the guardians after birth, is ''surrendering'' her own child, and treat the transaction as a particularly shady form of post-birth adoption. Instructive in this regard is the recent Michigan case, *Doe v. Kelley*.[9] The Does were husband and wife, Jane Doe incapable of bearing children. They proposed to have Mary Roe (who joined the complaint) conceive a child by John Doe through artificial insemination, carry it to term and deliver it to the Does. In return, Mary Roe would receive medical expenses, various forms of insurance, and significantly, $5000 for her trouble. Since the transaction was legally an adoption, this last provision brought Roe and the Does under the Michigan statute MCLA 710.54, which provides that ''Except for charges and fees approved by the court, a person shall not offer, give, or receive any money or other consideration or thing of value in connection with any of the following: (a) The placing of a child for adoption . . . "; another statute makes violation of the first a misdeanor, second offense a felony. Doe challenges the constitutionality of both on grounds of vagueness and violation of privacy. The judge dismissed the "vagueness" challenge—the statute is, after all, admirably clear—and took on the privacy challenge at some length, concluding that the statute was indeed constitutional and that the transaction contemplated by Roe and the Does was indeed illegal. The judge pointed out that the statute forbade not child-bearing or child-rearing, but a certain kind of monetary transaction, not covered by any right of "privacy." Even if it were covered, that right is not absolute, and the state's interest "to prevent commercialism from affecting a mother's decision to execute a consent to the adoption of her child" is powerful enough to overcome it. The "sale" or "bartering" of children, forbidden by law in all states, is "patently and necessarily injurious to the community." The judge dismisses the idea that no "adoption" would really take place, since the baby would really (by contract) be the child of the Does to begin with, quoting Holder at length:

> . . . The "foster" or gestating mother would presumably be considered by most courts the natural mother of the child since she and not the donor-mother was willing to go through the inconvenience, discomfort and dangers of pregnancy and childbirth.

> It is highly unlikely that a judge, faced with a conflict between two women, one of whom has delivered the child and the other of whom "should" have done so by normal means but who was too busy or disinterested, would resolve the issue of which is the true mother in any way other than by awarding parental status to the host-mother, contracts to the contrary notwithstanding. . . [10]

(There is certain irony in the judge's quoting a legal commentator predicting his action as the justification for the action itself. Law is a curious mixture of the empirical and normative.) The judge continues:

> The evils attendant to the mix of lucre and the adoption process are self-evident and the temptations of dealing in "money market babies" exist whether the parties be strangers or friends.

Voluntariness of the gestator in reaching the contract seems to be the reason for this condemnation: "How much money will it take for a particular mother's will to be overborne, and when does her decision turn from 'voluntary' to 'involuntary'."[11] That would seem to be a concern for the good will, good faith, or good intentions of the parties involved in the transaction. But the judge insists that the

> personal desires and intentions of plaintiffs are not in question, and their good faith is conceded. Nonetheless, public policy is established to guide all of the people of this State, of whatever intent.[12]

Not in this case alone, then, but in all cases of surrogate motherhood, the exchange of money shall be sufficient evidence that the sale of a baby has taken place in violation of law and in violation of fundamental human rights. No wonder that efforts to set up, or to support efforts to bring about passage of a statute that would render legal, such abhorrent arrangements, for amounts of money that seem to some to be coercively large, might seem morally as well as legally exploitative:

> The "rent-a-mother" possibility presents a basic problem of human exploitation totally unrelated to the morality of the fertilization process itself. It has already been established that some British physicians are prepared to pay the host-mother vast sums of money in order to induce her to carry another woman's child. For example, in 1970 a British embryologist at the University of Birmingham offered $4800 to any woman who was willing to gestate a baby for another woman. There was no followup story indicating whether or not he had any takers.[13]

And that was in 1970, before the last rounds of inflation. A similar objection is raised by the defendant prosecuting attorney in Doe v. Kelley (and quoted approvingly by the judge):

> Thus, contrary to plaintiff's exhortations, in all but the rarest of situations, the money plaintiffs seek to pay the "surrogate" mother is intended as an inducement for her to conceive a child she would not normally want to conceive, carry for nine months a child she would not normally want to carry, give birth to a child she would not normally want to give birth to and then, because of this monetary reward, relinquish her parental rights to a child that she bore.[14]

But morally, I submit, the situation does not appear anywhere near as bleak as the observations above would suggest. Holder raises two issues: the validity (or metaphysical possibility) of any contract to gestate a child for another, and the moral issue of exploitation of the gestator by payment for her services. As to the first issue, if we grant the metaphysical possibility that the unborn fetus is indeed a human being, then there seems to be nothing wrong with making appropriate arrangements for its loving care without changing its custody. I fail to see anything *morally* wrong with an agreement (presently, I concede, legally impossible) that would permit one woman to conceive and bear a child, or accept embryo transplatation of a child adopted at conception or *in utero* by another woman. In that case the prospective guardian is the parent of the child from the moment the adoption papers are complete, and the gestator is simply caring for another person's child, not growing her own with the prospect of giving it up in the future. As Grobstein suggests, "the role of wet nurse (which is accepted in many cultures) could probably be expanded fairly rapidly to include surrogate childbearing."[15] Surrogate motherhood becomes, on this understanding, an extension of existent child-care practices, not existent adoption practices. As such, it would be subject to all existent moral imperatives for contractually arranged childcare: the gestator would be under obligation to make reasonable provision for the child's health and welfare (i.e., in this case guarding her own health, avoiding violent activity, harmful substances, and so on), to have the child available for the parent to pick up at the agreed-upon time (at the birth of the baby), and above all, not to abandon or destroy the child. Holder's assumption[16] that the gestator would have a perfect legal right to abort a baby she had agreed to carry for another woman may be correct in law, but even the most ardent "pro-choice" partisan would have to be appalled at the moral character of such a violation of trust. Beyond these general strictures, I see no reason why statutes should not be amended to permit a contract fairly based on the mutual interests of the parties to it; that should be perfectly adequate to protect the rights of donor, gestator, and child.

With respect to the second issue, I find Holder's suggestion—that a payment of the "vast sum" of $4800 might be "exploitative" because of its large size—simply bewildering. Even corrected for inflation since 1970, that just is not very much money. And most day laborers, in fact most college professors, find low wages, where alternative employment is not available, much more exploitative than high wages. That, in fact, is the usually accepted definition of "exploitation": the use of a power position to arrange for the provision of services for oneself at a rate of compensation significantly *less* than what those services would be worth on an ideally open market. Presumably the rate of compensation should be set by each prospective gestator, in the light of the alternative employment open to her, the inconveniences and burdens of pregnancy for her personally, given her circumstances, and, eventually, the prices set by her competition. I see nothing wrong in paying a babysitter, and I see no reason to believe that this market would be inherently more exploitative than that for babysitters. As for the defendant prosecuting attorney's comment, that the nub of the evil of surrogation lay in the fact that the money served as an inducement to do things they otherwise would not do, it should be noted that the fact merely assimilates it to every other kind of employment. The defendant prosecuting attorney would not have made his speech in court but for the inducement of a paycheck, nor would the lawyer for the Does have appeared in court, nor would I prepare classes, set and correct examinations, and turn in grades for dozens of uninteresting freshmen, but for (ultimately) the inducement of a paycheck. No activity can be in worse moral shape, if done for pay, than the teaching of philosophy, whose Founder specifically enjoined against the taking of a fee for teaching.[17]

Possibly sociological assumptions are playing a disproportionate role in Holder's analysis:

> . . . it is also arguable that only a woman who was so poor that she had no other way to survive would deliberately rent her body for nine months of pregnancy followed by labor and delivery.[18]

For it is from this assumption mentioned repeatedly in the longer passage above, that she derives, first, a set of very practical medical difficulties attendant upon surrogate motherhood:

> If poverty becomes the motivation for an agreement to serve as a host-mother, serious medical problems for the baby resulting from the host-mother's preextant malnutrition, anemia and other problems could also be reasonably predicted to occur.

and second, the analogy with another class of desperate persons, the convicted criminals in our prison system who "receive more humane treatment as the result of their participation in dangerous research."[19] I like the analogy no better than the original case, but this is, at least, the reasoning that gives us the conclusion: "This situation, therefore, presents enormous ethical as well as legal difficulties."[20] But the sociological assumption does not seem to hold.

Anne Fleming includes in her article a brief survey of surrogate mothers currently working—i.e., pregnant or about to become so—in Louisville, Kentucky, where surrogation is (apparently) legal. They include two housewives, one secretary, one engineer, and a hospital worker. They must have children of their own to be eligible for this privately run program, and they do not seem to be particularly hungry. The hospital worker admits that giving up the baby will be the most difficult part, but she has two children, and wants no more of her own.

But having those children inside her and bringing them forth was the thing she did best in the world. So she wants to do it again. These do not seem to be complicated women and theirs is not a complicated impulse. "There are some some who just like to be pregnant. My mother had six children," one of them said, raising images of a fecund line. When she carried a child, this woman continued, she felt important in ways she otherwise never did. The idea of carrying someone else's baby makes her feel even more important because then she is not just harboring a child, she is harboring a gift.[21]

No medical dangers for the fetus seem concealed here; no parallel with frightened prisoners suggests itself. The dangers of "exploitation" seem, at least with the current crop of volunteer surrogates, to be over-stated. They are paid, of course, a fee, negotiable between $5000 and $15,000 ("You have to expect to pay more for an engineer"), but the fee hardly seems to be necessary for instant relief of poverty.

Exploitation of the gestator, then, does not seem to be as serious an ethical problem as Holder would have it. The legal problems however, are real, and will have to be dealt with; the fact that, given the contrary tradition of the law, very strict scrutiny will be given all statutes permissive of surrogate motherhood, should be further reassurance that the rights of all adult parties are protected.

The fourth issue at stake in surrogate motherhood raises an entirely different problem: that of the rights and moral status of the zygote. The supposition that the embryo is in fact a human being with an existence and rights of its own, hence eligible for adoption *in utero* (or even, in the case of in vitro fertilization, *ante utero*), may simplify the rights picture where gestator and prospective guardian are concerned, but complicates it for the physician or technician in charge of the fertilization, if it is to be done in the laboratory. Where possible, several ova are gathered to maximize the chances of a successful fertilization; when one is ready to implant, others are discarded; zygotes that fail to implant are simply lost.

The procedure seems inseparable from zygote loss. What is to be said of this? Are these really mini-abortions?[22]

If the embryo possesses a human existence "that makes protective claims on us," has rights of its own, do we not, having played God in creating it, now commit murder in discarding it?[23] Even if this is not properly called "murder," it could still be claimed that we violate the fetus' human dignity by manipulating it, experimenting with it with no possibility of consent, treating it as just one more laboratory substance.

Or from a consequentialist standpoint, it could be argued that we are
recklessly endangering the baby and the family in these procedures,
since we have no assurance at all that they will not result in hideously
deformed children (or abortions to prevent them).[24] In short, the laboratory procedures used in *in vitro* embryo development for implantation in the gestator may be sufficiently immoral that the laboratory
phase of the operation should be banned on ethical grounds.

Two approaches are possible here. One approach, strongly recommended by Grobstein,[25] simply denies that the "fertilized egg" (itself more of a process than a product) has any moral significance beyond that possessed by sperm and egg separately. The zygote is mere
tissue on this account, but tissue with an interesting property—that
properly inserted in a woman's body at the appropriate time of the
month, it will undergo a series of transformations that will eventuate in
a human being. The point at which this final stage is reached may be
subject to differing opinions: nidation, viability, and birth seem to be
the major candidates for that point at present, with viability command-
ing a probable majority of the votes. Whatever the critical point of becoming human, it is sometime *after* insertion into the womb of the
gestator, so the laboratory procedures never deal with what might
properly be called a human life.

There are difficulties with this approach. One obvious difficulty is
that not everyone will agree with the metaphysical premise—that individual human life does not begin until some point in time conveniently
removed from the manipulation of the fertilized egg in the laboratory.
Another is that it is inconsistent with the approach to surrogate gestation taken in the previous discussion. From the point of view of the
"defenders" of surrogate motherhood, viz., those who would like to
see research into these practices continued, and legislation permissive
of them passed, it is important to be able to present a unified account of
what happens in this kind of surrogation. If a significant number of the
electorate believe that an individual human life begins at conception,
and if that is the assumption used to explain the moral status of the
relationships among the adult participants during gestation and at birth,
it would behoove those engaged in this research to explain the laboratory procedures in the same way. But can the apparently callous disre-
gard of the possibility of introducing a defective fetus into the
gestator's uterus, and the routine disposal of unwanted embryos, be
reconciled with that assumption?

The answer seems to be that it can be, if we set a standard of respect for the embryo no higher than that already in effect in nature. For
the natural processes of conception, gestation, and birth are enor-

mously wasteful of fetal life; it has been estimated that up to 70% of natural *concepti,* eggs fertilized in the normal course of events, fail to implant or are sloughed off after a few weeks of gestation.[26] There is a particularly strong tendency for defective embryos to be sloughed off. The facts of normal development, then, provide two defenses for the experimenter: on the one hand, the practice of discarding living fetuses is only that employed by nature in the management of the natural economy of reproduction; on the other hand, any injury to the *conceptus* that would result in birth defects will probably be sufficient to ensure that the embryo will not implant or will not be retained in the womb, so the fear that defective babies will result from this process is probably unfounded.

The fifth area of concern is more difficult to comment upon than the others, for its issue seems to be precisely the unspeakable fears left over after analysis has resolved the problems. Let us recapitulate. Up to this point in this section, we have discovered, first, certain inefficiencies in the practice of surrogate motherhood—that is, if our object is simply to get babies into the hands of couples that want babies, it would probably be cheaper and easier to increase the number of babies available for adoption by compensating women for bringing fetuses to term rather than undergoing elective abortion. Second, we have discovered certain problems of justice in the allocation of public resources to research on *in vitro* fertilization, in the light of the prevalence of pressing health problems elsewhere, the very limited number of people who would be served by the new technology, the fact that the "health problem" that would be solved by it is by no means a life- or function-threatening problem, and the availability to those few people of alternatives to artificial fertilization (viz., adoption). All of these factors suggest that the money could best be spent elsewhere. Third, prescinding from formidable legal problems, ethical problems of exploitation of the mother foreseen by Holder and others were found to be less serious than projected, and probably no obstacle to the development of a reasonably supervised practice of surrogate motherhood. Fourth, ethical problems arising in the treatment of the embryo in the laboratory were found to be not immediately serious, given that the possible "abuses" of the embryo in the laboratory were no worse than the treatment of the pre-implantation embryo in nature. This is one area where we are not in a position, at present, to improve upon the standard that obtains in nature; if we ever are—if it becomes possible to rescue all zygotes and raise them (in, say, artificial wombs)—it will be time enough to consider what our obligations in this regard might be. Then what ethical problems are left? Are there areas of unexamined

consequences, or categories of rights and duties, that we have overlooked? Or are we talking only about a residue of nameless superstition
when we talk about the "moral essence" of motherhood and worry
about its violation?

For the same reason that natural motherhood has always fasci-
nated us, artificial motherhood has proved a fertile source of fictional
horror. From the unnatural creation of a single human "monster" in
Mary Godwin Shelley's *Frankenstein* to the unnatural manufacture of
thousands of identical babies in Aldous Huxley's *Brave New World*,
writers have capitalized on the certain revulsion producible by mention
of such practices. (Interestingly, the liquid in which Huxley bathed his
embryos to nourish them, modifiable at the manufacturer's will to produce different classes of citizens, was also called "the surrogate.").
Nor is such automatic horror gone today. Holder mentions, at the outset of her much more recent discussion, objections that

> range from the assertion that research in this area violates the
> "will of God" and dehumanizes people to fears that once artifi-
> cial fertilization and gestation are possible, governments will cre-
> ate hundreds of beings with which they can control either their
> own citizenry or the world.[27]

These lie at the heart of the terror we took at the opening of this paper;
but on closer inspection, there is no "range" between the assertions
she cites: both assume that Motherhood belongs to God and Nature;
that once wrested from them and given over to human control, all
safety is gone. Control will quickly pass from the well-intentioned fertility specialists to the dictators, who will use the technology for their
own terrible purposes. Since the latter technology is as remote as it is
needless (naturally born human beings are malleable enough for any
dictator), and since the assumption of any inevitable "slippery slope"
from good to evil practices in this area is undefended and unjustified,
these objections can be taken to exemplify the superstitious side of the
opposition.

Specific religious traditions can, of course, bring objections to
bear on surrogate motherhood and all allied practices from the perspective of a unique body of doctrines. Roman Catholics, for example,
have grave difficulties with the discarding of embryos in particular, but
also with the whole range of conception-assisting technology in all its
aspects.

> Roman Catholics and other opponents of the conception revolu-
> tion believe that sexual intercourse is the only "natural", hence

ethically acceptable, form of conception; that an intrusion by a third party—be it a sperm donor, a surrogate mother or a test-tube scientist—in the act of procreation is destructive of the institution of marriage.[28]

Hellegers and McCormick put the case more strongly:

> And, in time, the next step may be the surrogate or host womb, where a woman, from medical necessity or convenience, cannot or will not bear her own child. The embryo transfer and child will be carried by a third party. This may seem incredible, but today's incredibles are often tomorrow's facts, and it is the part of wisdom to recognize this. We see in these procedures grave assaults on marriage and the family, to say nothing of the subtle devaluation of sexual intimacy that clings to them. We see, therefore, serious, indeed insuperable, moral objections in such developments.[29]

There is, to the best of my knowledge, no empirical evidence that the employment of some form of reproductive technology actually results in a higher divorce rate. I should imagine that the statistics would indicate that marriages in which the desire for children is strong enough for the couple to seek out such techniques are more stable than the average. So the "assault on" and "destruction" of the institution of marriage referred to in the passages is not an empirical construct, a change in facts about the marital relationship, but an analytic one—no matter what the parties to the marriage may think or feel on the matter, no matter what the effects of the baby on the family, where artificial means of conception or gestation have been introduced, the *real* marriage is ontologically damaged. Such definitions hold only within the religious tradition whose views are in question, and cannot be expected to influence opinion, or even make sense, outside them.

A more general opposition to reproductive technologies, not conditioned on adoption of a certain set of religious doctrines, is expressed by Paul Ramsey:

> I'd rather every child were born illegitimate than for one to be manufactured . . . Already women think of themselves as machines of reproduction. Look at the ease with which young girls have abortions, so sure that they can have another child any time they want. And now women are selling their bodies for nine months and people are talking about freezing fertilized eggs. Pretty soon, a woman will be able to go to the supermarket and pick out an embryo.[30]

Ramsey's choice of words alerts us to the presence of the Promethean terror in its pure antitechnological form: a child whose creation is aided by the new reproductive techniques is "manufactured"; his mother already sees herself as a "machine." Young girls assume they can make babies "any time they want." Bodies are for sale, fertilized eggs can be frozen, embryos are to be picked out, freely chosen, so the supermarket—presumably the frozen goods section—the 20th century home of unlimited free choice, of anything you want, is the appropriate place to market the bodies of babies. Throughout this very revealing passage, Ramsey contraposes the God-given mystery of pregnancy and all the images of Man the Technologist serving Man the Consumer— the machines that go on and off at the touch of a button, the vast supermarket where everything is processed, preserved and packaged for sale, attractively displayed on shelves to be chosen or left, precisely as you want, any time you want. The place of this voluntarist lifestyle in the moral order is deftly indicated in the first sentence; it is worse than illegitimacy; illegitimacy is seriously immoral (the result of incontinence, and in violation of the reproductive monopoly of the institution of the family); therefore this lifestyle is *very* seriously immoral, and should be avoided, and its practices condemned and contained.

But on closer inspection, what is there of argument in the passage, beyond the contraposed images? There seems at one point to be a consequentialist argument, in the form of a dark prediction that numerous abortions may lead to difficulties in conceiving or bearing a child later on; but this is not really said, and it does not seem to be true in any case. There seems at another point to be a deontological argument, in characterizing surrogate gestation as "selling . ... bodies", like prostitution, which is forbidden, on an analogy with slavery, as a violation of human dignity. But he does not argue this point either (as Holder does), and as we have seen, other nonexploitative analogies are available for surrogate gestation. In the last analysis, there is no argument here—or rather, the image is the argument—and we shall have to be content with that.

It should not come as a surprise that a significant portion of our national dialog is carried on in poetic images rather than logical arguments. For Ramsey, the appropriate images to be contraposed are that of "natural" pregnancy (to be approved) and "the machine" or "manufactured" pregnancy (to be condemned). Earlier we saw that the practicing surrogate had come up with a different image for surrogate gestation: "she is not just harboring a child, she is harboring a gift." And the attendant publicity generally underlines the generosity of these women: "their gallantry shines back at them from the television set."[31] Ramsey does not have a monopoly on the public use of symbols. Ultimately, whether his mechanical imagery or the surrogates' philanthropic imagery prevails for the description of surrogate motherhood depends on the outcomes of the other moral problems we have noted: if abuse and exploitation run rampant in our early experiences of the practice, once we have learned to do it successfully, if medically implanted babies are regularly defective, if disproportionate amounts of money are consumed in this research to no avail while born children starve, then the whole practice will have to be abandoned and Ramsey's images will be used to explain why. If the abuses are avoided and the practice becomes common, the surrogates' images will be used to recommend it. Eventually, in short, the images we entertain of this or any social practice will match the policy we have enacted toward it.

Motherhood in Human Perspective

The problem we began with was the presence in the society of a new technology of laboratory-assisted human reproduction, and the presence in the society of a deep-seated distrust of that technology. We set out to discover whether the new techniques are Aesculapian in their beneficent powers, as the proponents aver, or Promethean in the violence they do in provinces not appropriate to human technology. It is not clear that we have answered the question; indeed, it is reasonably clear that we have not. The failure can be attributed in part to the stage of the subject matter—the technology is young and changing rapidly, very little serious work has been done on the subject. But the failure occurs primarily because of a simple divergence of evaluation: whether the technology is seen as good or bad depends upon which set of larger social norms it is measured against; and a refractory battle of images takes over in the last resort. This essay has been, of necessity, exploratory in nature. But I would like to advance the following statements as loosely derivable from the exploration to this point, not as conclusions, but as way-stations in the continuing discussion:

1. It is time to demythologize reproductive research. Let us admit that all medical technology "tinkers with human life itself," that technology in general has been around much too long for us to continue viewing it with Victorian alarm, that dark mutterings about "provinces of the gods" are no longer the privilege of a nation that sends its

rocketships about the universe. The poetic imagery meant to arouse our terror and digust must be taken as empty unless backed up with specific ethical arguments. Neither Aesculapius nor Prometheus, but human beings, must answer to God (and to each other) for the use of these techniques.

2. Stripped of quasireligious myths, the techniques of *in vitro* fertilization seem to pose no insuperable moral problems. The ethical constraints on research and practice in this area need not, apparently, differ from those on biological laboratory and medical personnel generally.

3. No grave problems of exploitation arise of necessity in contractual arrangements for egg donation or for surrogate gestation. Egg donation can best be understood on an analogy with sperm donation, surrogate gestation on an analogy with surrogate care of a born child. In contracts for surrogate gestation, as in all contracts for services over a long term, the possibility for coercion exists, abundant possibility for changes of heart exist, and unforeseen circumstances may bring to light large gaps in the mutual understandings that the signatories held at the signing of the contract. But it is not clear that present legal provisions for review and modification of contracts would have to be supplemented to handle contracts of this kind, but for the pervasive (and, it could be argued, perverse) inclination of the courts to treat them as contracts for the sale of babies. As it is, statutory relief will be necessary.

4. Difficulties raised under the headings of justice and efficiency are more troublesome for the new reproductive technologies. There is no dearth of people in the world. There is no social need at all, save on eugenic grounds that can be shown to be spurious, to reproduce people who cannot reproduce themselves. We are catering, in this technology, to the desires of a very few people, and the fact that the desires are very deeply felt by those people must be weighed against alternative uses for those medical resources, on the one hand, and on the other hand, against the relative ease of procuring babies by other methods (importation from Third World nations, cash payments to pregnant women considering abortion, and so on).

My own position is that the research should continue, as all research should continue, to find out what can be found out. Possibly the most important next steps to be taken are in the law, to ensure that surrogate gestation arrangements, already being concluded informally, contain adequate protections for all parties involved.

Acknowledgments

I would like to acknowledge the help and encouragement of Angela R. Holder and Robert J. Levine, both of whom read an earlier draft of this article and made many helpful suggestions. The errors that remain are my own.

Notes and References

[1]Angela Roddey Holder, *Legal Issues in Pediatrics and Adolescent Medicine* (New York: Wiley, 1977). Michael H. Shapiro and Roy G. Spece, Jr., *Cases, Materials and Problems in Bioethics and Law,* American Casebook Series (St. Paul, Minnesota: West Publishing Co., 1981). There is also a recent informal treatment of the topic by a lawyer active in setting up surrogate pregnancies: Noel P. Keane, with Dennis L. Breo, *The Surrogate Mother* (Everest House, June 1981). The major recent source is Anne Taylor Fleming, "The New Conception" (The New York Times Magazine section: July 20, 1980). Two other serious reactions to the birth of Louise Joy Brown (the first test-tube baby) contain precisely one paragraph each on the topic of surrogate motherhood: Andrew Hellegers and Richard McCormick, "Unanswered Questions on Test Tube Life" *America,* August 19, 1978, pp. 74–78, and Clifford Grobstein, "External Human Fertilization," *Scientific American* 240 (June 1979), 57–67.

[2]As Landrum B. Shettles, a prominent reproductive scientist at the time, put the case, "If the bridge is out, what's wrong with using a helicopter?" See David Rorvik, "The Embryo Sweepstakes" (The New York Times Magazine section: Sept. 15, 1974) 17ff. Cited Holder, *op. cit.* p. 4. Since his participation in the notorious DelZio case (see DelZio v. The Presbyterian Hospital, United States District Court, Southern District of New York, 1978 74 Civ. 3588) Dr. Shettles appears to have left this field. In the case of Louise Brown, of course, the husband was the donor. Were fatherhood the focus of the paper, a distinction would have to be drawn between the moral implications of Artificial Insemination by the husband (AIH) and AID; such is not our present concern.

[3]See for example, *Exodus* 20: 14, *Deuteronomy* 5: 18.

[4]See for example, *Leviticus* 20: 10, *Deuteronomy* 22: 22–24.

[5]Holder, *op. cit.,* p. 12.

[6]David Rorvik, "Embryo Transplants" (*Good Housekeeping,* June 1975) p. 78, Cited Holder, p. 5.

[7]See Holder, pp. 5–9.

[8]Holder, p. 7

[9]Circuit Court of Wayne County, Michigan, 1980. Reported in 1980 Report On Human Reproduction and Law II A-1. Exerpted in Shapiro and Spece, *op. cit.* pp. 537–542.

[10]Holder, *op cit.*, p. 7; cited Shapiro and Spece, p. 540.

[11]Shapiro and Spece, *loc. cit.*

[12]Shapiro and Spece, p. 541.

[13]Holder, p. 5.

[14]Shapiro and Spece, p. 541.

[15]Grobstein, *op. cit.*, p. 63.

[16]Holder, *op. cit.*, p. 9.

[17]Plato, *Apology*.

[18]Holder, *op. cit.*, p. 6.

[19]*Loc. cit.*

[20]*Loc. cit.*

[21]Fleming, *op. cit.*

[22]Hellegers and McCormick, *op. cit.*, p. 76.

[23]Cf. the sins of Daniele Petrucci (1959), cited Holder, *op. cit.*, p. 2.

[24]Hellegers and McCormick, *loc. cit.*

[25]Grobstein, *op. cit.*, pp. 64–65.

[26]Grobstein, *op. cit.*, p. 61.

[27]Holder, *op. cit.*, p. 1.

[28]Fleming, *op. cit.*

[29]Hellegers and McCormick, *op. cit.*, p. 77.

[30]*Loc. cit.*

[31]*Ibid*, p. 23.

Section III

The Distribution of Health Care

Introduction

In "Scarcity and Basic Medical Care" Robert Almeder confronts the problems of increasing scarcity in basic health care services and sidesteps the question of whether the government should intervene forcefully in the health-care marketplace to provide more health care for the poor and the elderly. Rather, he believes that since the government *will* so intervene, the important question is not *whether*, but *how*, that intervention should proceed. His essay seeks to answer this latter question. In seeking the answer, he urges that the government should not seek to decrease the scarcity in basic health services by manipulating the demand for such services. Instead, the government should intervene to increase the supply of basic services. After spelling out in detail how the government should increase the supply of basic health services, he defends his proposal against a number of predictable objections. As the author sees it, his proposal has the virtue of being effective, while mediating between total socialization of the health care industry and total abandonment of the problem of scarcity to the mechanisms of the free-market economy.

In "Distributing Health Care: A Case Study," Professor Nicholas Fotion approaches the problem of scarcity in a very practical way. He begins with a detailed case study of how one particular community, Grandview, confronted the problem of scarcity. His tentative conclusion is that a careful examination of the problem warrants the belief that the problems of scarcity are more likely to be solved by the free market than by anything the government could do by way of intervention. To the extent that most communities are like Grandview, he urges that government intervention is more likely to exacerbate the problem than it is to solve it. Near the end of the paper, Professor Fotion seeks to defend the mixed system of health care delivery. In this latter effort he confronts other recommendations and defends his position against other proposals.

Both essays are distinctly pragmatic in tone and avoid philosophical abstractions or philosophically ideal, rather than currently workable, solutions.

Scarcity and
Basic Medical Care

Robert Almeder

Introduction

The seriousness of the ever-increasing scarcity of basic health care in America frequently prompts the following question. Should the government forcefully intervene in the medical health-care delivery system in order to guarantee that all citizens receive basic medical services, or should the government stay out of the health-care delivery system and thereby allow the mechanisms of the free market to distribute health care like any other service under the formula of supply and demand? As things presently stand, the government has intervened via Medicare and Medicaid to provide health services for the poor and the elderly. But the problem is that in inflationary times the government must put a limit on the amount of Medicare and Medicaid dollars thereby leaving a number of people with less basic care or none at all.

Those who think the government should intervene forcefully usually defend their preference by arguing that in a democratic society every citizen has a natural right, or some sort of a basic right, to affordable basic health services. Without one's health, they say, one does not have equal opportunity to the benefits of society. They then urge that since it is the purpose of a democratic government to protect the basic rights of its citizens, it should, by implication, guarantee the basic health of its citizens.[1] And there are other arguments supporting the view that health care is a basic right in a democratic society.[2] For those who accept this basic line of reasoning, the question is *how,* and not *whether,* the government should intervene forcefully in the medical marketplace. Of course they would not hold this view if they felt that the medical marketplace, left to its own devices, could well provide for the basic health needs of all at affordable prices. But that is just what they see as unlikely to ever occur in a free medical marketplace. Presumably it was just this sort of thinking that gave rise to the National Health Service. And naturally there are some fairly radical positions that seek the total removal of the health care system from the marketplace so that everybody is treated according to the same standard of health care.[3]

Alternatively, those who favor keeping the government out of the marketplace, and out of the health care business, generally do so by attacking the view that everybody in a democratic society has a natural right to health care service. In the opinion of this latter group, while it may be shameful, it is certainly not unjust, to let people suffer who cannot afford basic health care services.[4] On this view, one's health is one's responsibility or duty; but one does not have a right to it. This latter group may even urge that in the long run the best interests of society as a whole will best be served by allowing the marketplace to operate freely even at the short-term expense of losing some people to bad health, people who cannot afford the fee for the health service needed.

Sometimes, too, there are very pragmatic arguments for keeping the government out of the health care business. For example, we are occasionally told that government intervention in the health care system just is not necessary; the health market left to its own devices will quite nicely provide basic health care for all those who need it. And we are also told that even if the medical marketplace fails to provide everybody with basic health care, still, there is nothing so bad that cannot be made worse by letting the government do it. As the saying goes: if you like the postal service, you will love socialized medicine.

Bypassing for now the question of whether a free medical marketplace can provide for the basic medical needs of all, the two opposing views just discussed admit of subtle and prolonged philosophical discussion. Moreover, it would appear that, pragmatic considerations aside, the proper philosophical answer to our opening question may well hinge on just what we decide is a basic human right. On this score, we can note that philosophers are by no means agreed and that there are long-standing philosophical debates over the nature of human rights. Indeed, it is fair to say that there are radically different philosophical conceptions of what a basic human right may be; and there are some philosophers who think that the debate over the proper characterization of human rights may well have no objective answer.[5] Furthermore, for those who are not versed in philosophical analysis such discussions are not only irrelevant (because they are incomprehensible to those who are not philosophers, and sometimes even to those who are), but also exasperating because they convey the impression of fiddling while Rome is burning. Accordingly, it may well appear the better part of philosophical valor and prudence to ignore the opening question because it is unlikely that anything much will come of trying to answer it.

In this essay, then, I propose to answer what is seemingly a more modest question, but one that needs to be asked and at least seems to be answerable without an interminable philosophical discussion on the nature of human rights and distributive justice. The question is this. Assuming that the government *will* intervene forcefully in the basic health-care delivery system to decrease the scarcity in basic health care services, what would be the most philosophically, politically, and economically effective mode of intervention? For the sake of discussion, then, I shall assume what I think is true (and for which I shall later argue), namely, that sooner or later our government will need to set a clear limit to the amount of health dollars it will spend for basic health care. Thereafter it will be forced to enter into the health care market to assume a decisive role in making more health care available at a reasonable cost for all its citizens. Some of the reasons for thinking this assumption is true should emerge as we progress.

After proposing an answer to our philosophically more modest question, I shall consider some objections to it and offer a few refinements of it. But the fundamental point here is to offer a timely and workable, rather than an enduring and philsophically ideal, proposal for decreasing the scarcity of basic health services in America *via* governmental intervention.

Eliminating the Scarcity

Cutting the Demand

Most economists agree that, in a free marketplace, if the demand for a product increases and the supply remains constant, competition for the product drives the price up proportionately to the demand. In the case of basic medical services in America, we seem to have an ever increasing demand for basic medical services owing to (a) geometrical increase in population and (b) basic changes in lifestyle generating a higher incidence of stress-related ailments and cancers. Naturally, the increase in demand would not produce the scarcity of the product (and the corresponding cost increase) unless there were also a failure to increase the supply proportionately to the demand. So, from the viewpoint of standard economic theory, the quick solution for alleviating the scarcity of basic medical services is obviously to either (a) dramatically decrease the demand for basic health services, or (b) increase the supply to meet escalating demand, or (c) both decrease the demand and increase the supply proportionately.

Now let us suppose, for the sake of discussion, that what we can do is not in any way limited by available funding. Of course this is a false supposition; but let us suppose it anyway just to make a point. Given this supposition, then, if we decide to try to drive down demand for basic health services, we could do so by seeking to control the population. Otherwise, on the principle that a *significant* (if not large) proportion of the demand can be eliminated by teaching people the virtues of avoiding stress-related ailments (or ailments produced by poor nutritional behavior) we could begin a massive preventive medicine program. Or we might try to do both.

Interestingly enough, however, even if we could institute population control measures (a politically unlikely occurrence in contemporary America), and even if we could successfully mount a massive preventive medicine program (an equally unlikely occurrence given the scarcity of funds combined with priorities oriented toward medical cures rather than prevention of disease), there is no guarantee (and some would say very little likelihood) that the scarcity in basic health services would be diminished very much. Why? Well, given a constant supply, decrease in demand will tend to make the product less scarce (and less expensive) only if there is *competition* (or better yet, only to the degree that there is competition) in the marketplace. And there is very little competition in the medical marketplace. Let me explain.

In a marketplace that is not free, pricing of the product (and by implication scarcity) is *not* a function of the relationship between supply and demand. For example, suppose (again just for the sake of discussion) that the supply of physicians and hospital services remains constant and we are able to drive the demand for those services down at least 50%. As long as the demand still outstrips the supply, there is nothing to stop physicians and hospitals from raising the price of services 50% to offset the decrease in demand and thereby maintain a constant level of income. Indeed, as long as hospitals function as corporate entities with responsibilities to shareholders, and as long as physicians are reluctant to work longer and harder to maintain the same level of income (and as long as most costs will be paid by the government and third party payments), it would be naive to expect a decrease in scarcity of medical services because it is naive to expect a decrease in overall cost for an ever decreasing number of basic services. As a matter of fact, K. Carney has noted that decreasing the number of hospital beds, for example, in an effort to contain hospital costs, can generally be expected to increase the cost of the remaining beds so that the cost of hospitalization will increase (or remain the same) even when the demand for beds is cut.[6]

Accordingly, as long as the supply of medical services remains constant, a decrease in demand for the services will tend to eliminate scarcity and drive down the price for such services only if the market is competitive. But as long as the price can escalate independently of demand (as it does when, whatever the cost, it will be paid mostly by the government or some third party), there would be no competition and it would make no difference how far down we drive the demand. However, if we put ceilings on costs for third party payments and government payments, then a dramatic decrease in demand would stimulate competition and tend to drive prices down. But this is to say that manipulating demand will be effective as a means of driving down prices and decreasing scarcity only if we are committed to price controls on such services. Without price controls, then, limiting demand for health services will not increase the supply of such services as readily as it would increase the cost of the ever decreasing services. Therefore, because manipulating demand will not tend to decrease scarcity and drive down prices unless we impose some price controls, it is unlikely that we can be successful simply because there will be so much resistance to the imposition of price controls. If possible, we should avoid price controls in an open market economy. And even if we could succeed without price controls here, it is questionable that we would ever be able to drive down the population or implement a meaningful program of preventive medicine for all the reasons mentioned earlier.

Naturally, there is ample evidence of strong forces working on the marketplace to keep the supply such that the demand, however it is manipulated, will considerably outstrip the supply. The American Medical Association, for example, resists any moves that would increase the number of physicians available for basic health care delivery. As a matter of fact, The American Medical Association currently claims that there is a significant oversupply of physicians although, to be sure, physicians generally defend their high incomes on the grounds that they must work harder and longer than most other professionals.[7] If there were an oversupply of physicians, and if the market were truly competitive or free, we should expect in virtue of the glut of physicians a decrease in the cost of basic services. The fact that basic medical services (office visits, surgery, testing, examinations, consultations) have escalated considerably in excess of the annual inflation rate, is evidence that if there is a glut of physicians, the pricing of their services is proceeding quite independently of supply.

In conclusion we are not likely to succeed in eliminating scarcity simply by decreasing the demand for services. Even if we could significantly decrease demand, which is very unlikely, we would need to impose long-standing price controls in order to make sure that pricing does not proceed independently of demand. Given the difficulties involved in decreasing demand, along with the inherent repugnance of price controls in an open market economy, we should turn our attention to the possibility of decreasing scarcity by increasing the supply of medical services.

Increasing the Supply

Unless there is a dramatic increase in the supply of physicians and hospitals, combined with appropriate and temporary limits to government and third party payments, gestures in the direction of increasing the supply will be of little avail in terms of hard results. Moreover, in terms of public policy, workability, cost, and infringement in the marketplace, this way of increasing the supply of medical services (and thereby limiting the scarcity in such services) seems most promising because it is politically workable, does not require permanent price controls, and because we can more readily increase the supply of basic services than we can diminish present demand for those services.

After all, given that anything we *can* do is always limited by available funding, and given the immediate outlay required for an effective preventive medicine program, it would surely cost the government considerably less to dramatically increase the supply of physicans than it would to drive down the demand. Such an increase in supply could be achieved simply by tying the existing federal outlays to medical schools to appropriate increases in the number of physicians graduated by our medical schools and by providing financial incentives (by way of federal subsidy) to those medical schools graduating larger numbers of physicians. This would seem to be the most cost effective method of generating an important increase in the population of physicians. Provided we place a dollar limit on government and third party payments, this would seem to be the most desirable method for driving down scarcity on health care services. As a matter of fact, given the geometric increase in the population at large, we should tie all federal funding for medical schools to corresponding increases in the physician population, increases that outstrip the predictable increases in the general population, by 20% annually. Until the price for general medical services drops to where the median physician's (and general hospital administrator's) annual income compares favorably with a well-paid public servant (say a senator, a congressman, or a federal judge), the price of medical services should be regulated by the government and indexed to the inflation rate. Thereafter, we should reexamine the government's role in regulating the supply side of the formula. But the regulation of prices here will be quite temporary. In the meantime, until the supply side of the formula is corrected, indexing the rate of basic medical costs to the annual inflation rate seems quite desirable as a temporary move.

If it be objected that this proposal is unfair because it limits the income of physcians and other health care distributors, the reply is that the action is justified as an indirect effect of the need for any form of government intervention for the purpose of driving down medical costs. Efforts at cost containment that focus exclusively on limiting demand will, for the reasons noted, be quite inadequate for driving down costs.

This proposed way of limiting the cost of basic medical services will meet stiff resistance. But its overriding virtue is that eventually the general cost of the government health budget could be reoriented for a national emphasis on preventive medicine that I see as our ultimate priority, but one that will never materialize unless we drive down dramatically the general cost of basic health care.

In the end, the attractive features of our proposal are that it is relatively simple, easy to understand, inexpensive to implement, likely to be effective, and politically capable of immediate implementation. It also mediates between total nationalization of health services and total abandonment of the health care delivery system to the mechanisms of the free market. Nationalization of health services is just a bit too untimely for our present capitalistic economy, and abandoning the health care system to the free market is just a bit too insensitive to the health care needs of most people, especially the poor and the elderly who would suffer most under that proposal. Of course, the unattractive feature of the proposal is that it recommends temporary price controls until the increase in supply of health services drops the cost. It is possible of course that we could drop the temporary price controls and simply put ceilings on government and third party payments. But that might beget unwanted and steep increases in cost on the short-term basis.

Objections

There are some fairly predictable objections to all this. For example, ardent supporters of the free market will urge that the government leave everything to the mechanisms of the free market and simply pay the cost for the poor, the elderly, and all others who cannot afford the usual fee for the service. Well, of course, the above proposal is predicated on the assumption that the government *will* intervene more forcefully in the health care delivery system as a result of the society's need to put a limit on the amount the state will pay for third party payments. As long as the government would assume the cost for Medicare and Medicaid, there would be no problem. However, when the government starts to limit dramatically the allocaton for health care, as it seems to be doing now, either basic health services will be dramatically curtailed for those who cannot afford it (and expensive for those who can), or the government will need to intervene in the marketplace and, by driving the price down for such services, make them available for those who could not otherwise afford them. Left to its own devices, and with the government limiting the funds for health care, the free market has no mechanism for providing the poor, the elderly and the unemployed with basic health services. But of course that in itself is no reason to nationalize health service. Naturally, if the government is not forced to intervene, then my proposal will make no sense. We are not arguing that the government *should* intevene, but only that if it must, the way here proposed is the most desirable for the various reasons stated above.

It might be further objected that our proposal is a bit too narrow and unfair to physicians because it only seeks to decrease physician costs and not hospitalization costs, the latter of which constitutes a major factor in escalating medical costs. Well, of course, this is true. It is equally true that simply increasing the supply of hospital beds will not drive down the cost of hospitalization, rather than push it up, owing to the expensive equipment necessary for running a hospital. So, our proposal ought to be supplemented with the recommendation that the government continue in its effort to induce competition in hospitalization costs by actively providing a favorable climate for more Health Maintenance Organizations (HMOs). In many cases HMOs have been quite effective in inducing competition among hospitals and their use holds out the prospect of some continued success as a means of driving down hospitalization costs.

Whether the overall effect of widespread use of HMOs will have a significant cost impact is not terribly clear. If they do, then the government may not need to intervene and regulate rates. But it is by no means clear that physician costs are any more competitive as a result of the widespread use of HMOs. At any rate, if the government is not able to stimulate sufficient competition in the medical marketplace by fostering incentives for widespread use of HMOs, then it will need to act

more forcefully to regulate the price and increase the supply of physicians and hospitals.

Finally, there are some who would object that this proposal is
much too weak because it underestimates the forces that will seek to
prevent a dramatic increase in the supply of basic health care services.
This sort of objection urges that even if we dramatically increase the
supply of physicians, still, as long as the government does not set the
price for health care services, the cost of those services will not go
down because there will be a concerted effort to keep the price of basic
health services artifically inflated, with the strongest members of the
health care community doing what is necessary to keep the prices up
no matter what the demand. To this objection the following reply is in
order. Our proposal requires that there be a *dramatic* increase in the
supply of physicians and that the increase be monitored annually in
terms of increased graduations from medical schools. Whether the
forces in the marketplace can succeed in keeping the price of basic
health services high in the presence of a long-term increase in the supply of physicians is a matter of conjecture. But if it were to happen, it
would constitute good grounds for more forceful regulation to achieve
the end for which intervention was prompted. To the extent that the
government is incapable of bringing competition into the health care
marketplace, then the health service industry will be stronger than the
government and monopolistic in its activities. At the moment, however, there would not seem to be any justification for thinking more
strenuous measures ought to be employed and prices set by government. The extreme measure of direct price control should not be considered before our proposal, which allows for less interference in the
marketplace.

Conclusions

The proposal made here is simple. If we must intervene in the health
care delivery system to drive down costs and allow affordable health
care for all, we should do it in a way that is effective and maximally
consistent with the principles of a capitalistic democracy. I am urging
only that we do not try to decrease the scarcity by manipulating the
demand. Rather, if we must intervene, we should do so by increasing
the supply of the product. In this age of supply-side economics, what
could be nicer? This is a fairly conservative proposal and one that may
well succeed if the supply is dramatically increased to the point of real
competition. Should it fail, then we would be driven to the conclusion

that if the government must (or will) enter into the health care system (rather than simply give the medical industry a blank check at the expense of the taxpayer), it will probably need to control prices on a regular basis or train physicians to provide health services at government prices. That would surely provide competition in the marketplace.

Effective administering of this proposal would require administrative insight and if the government were to give the National Health Service whatever powers it needed to succeed with such a proposal, I believe this country could move in time into the area of preventive medicine and thereby provide the civil body politic with the best possible health service available.

Finally, we have said nothing here about nonbasic medical care, the kind of care that attends exotic medical lifesaving therapy or long-term costly treatment for chronic ailments. It seems to me that there will always be scarcities in these areas for a nation trying to provide basic health care at reasonable prices for all. In this regard we shall need to develop a decision procedure to allocate such care. But more on that later.[9]

Notes and References

[1]This line of argument is developed by N. Daniels in his "Health Care Needs and Distributive Justice", *Philosophy and Public Affairs* 10 (1981), 146–179.

[2]See, for example, Baruch Brody's recent 'libertarian' defense of the basic right to health care in his "Health Care for the Haves and the Have-nots: Toward a Just Basis of Distribution" in Earl Shelp, (ed.), *Justice and Health Care,* (Dordrecht, Holland: D. Reidel Publishing Company, 1981), pp. 151ff.; and Charles Fried's "Equality and Rights in Medical Care", in J.G. Perpich (ed.), *Implications of Guaranteeing Medical Care* Washington, DC: National Academy of Science, Institute of Medicine, 1981, pp. 3–14.

[3]In reply to Kenneth Arrow's "Uncertainty and the Welfare Economics of Medical Welfare", *American Economic Review* 53 (1963), 941–973, Norman Daniels urges in "Health Care Needs and Distributive Justice" (see note 1 above) that health care services should not be marketed at all, even in an ideal market.

[4]See, for example, H. Tristram Englehardt Jr. "Health Care Allocation: Responses to the Unjust, the Unfortunate and the Undesirable," in Earl Shelp (ed.), *Justice and Health Care,* Dordrecht, Holland: D. Reidel Publishing Corporation, 1981, pp. 121 ff.; and Leon Kass, "Regarding the End of Medicine and the Pursuit of Health", *Public Interest* 40 (Summer, 1981), 11–42.

[5]This is because the two basic views about rights reduce to two mutually exclusive views, namely, *teleological* (under which a right is to be judged in terms of what is in the interest of the greatest good for the greatest number)

and *deontological* (under which a right cannot be set aside in the interest of the greatest good for the greatest number). Some philosophers continue to believe that there is no objective way to adjudicate the difference of opinion between these two basic positions.

[6]See Kim Carney,"Cost Containment and Justice", in Earl Shelp (ed.), *Justice and Health Care*, (Dordrecht, Holland: D. Reidel Publishing Co., 1981), p. 173. See also Salkever and Bice, "Hospital Certificate of Need Controls", *AEI for Public Policy Research*, Washington, D.C., 1979.

[7]For example, the Graduate Medical Education National Advisory Committee recently submitted to the Secretary of Health and Human Services a report arguing that there will be an oversupply of 70,000 physicians in America by 1990 and that dramatic action should be taken to limit the number of new physicians. See *Anlage* (Winter, 1981), 11.

[8]See Jon B. Christianson, "Can Business Stimulate Competition in the Health Care System?," *Business and Society* 19(2) and 20(1) (1980), p. 15 ff.

[9]See my "On the Role of Moral Considerations in the Allocation of Exotic Medical Lifesaving Therapy," in James M. Humber and Robert Almeder (eds.), *Biomedical Ethics and the Law* (New York: Plenum Publishing Corp., 1978).

Distributing Health Care

A Case Study

N. Fotion

Grandview's Story

This article says what it says primarily by way of illustration rather than demonstration. By extensively illustrating a case study of one community, it points to some of the complexities facing contemporary American society in its efforts to distribute health care. Because it is illustrative, the conclusions and recommendations found in this article are suggestive only. One would hardly want to draw any hard and fast conclusions from a sample of one (community). Nonetheless, the case study approach has some advantages. It avoids the error, commonly made by philosophers and theologians, of applying abstract and partisan principles to problems whose existence, seriousness, and nature have as yet to be determined. It also avoids the error commonly made by those who are more empirically oriented of studying one phenomenon (e.g., total health-care expenditures) in many communities and thereby missing out completely on how one fact on the health-care scene impinges upon another.

Since the case study approach requires that description precede explanation, the first half of this article is devoted to a description of the health-care scene in a community that I shall call Grandview.[1] The second half is divided in two parts. The first of these parts is devoted to making some general comments about the lessons to be learned from this description; the second to defending a position about how health-care should be distributed.

Before telling Grandview's story, some preliminary comments are in order. The background data reported in this paper were gathered in 1978 by means of telephone interviews (400 residents of Grandview and 200 in the surrounding area), two focus groups (composed of 15 residents each), personal interviews of hospital administrators and physicians (30 and 35, respectively) from Grandview and the outlying areas. I conducted all the personal interviews in the capacity of research director of the study. As will be related shortly, starting in 1979 my role as researcher changed to that of consultant.[2]

Grandview is an old, blue-collar river town of about 60,000 residents. It is not particularly affluent, although extensive poverty is not common either. ABC Industries, the largest employer in town, with approximately 6000 employees, produces heavy farm equipment. The union at ABC is strong and has used its strength to give its members good wages and a health benefits package that even includes dental care. Following the lead of ABC, other unions have helped their members achieve reasonable wages and health-care coverage. What with employers helping to provide health-care insurance to some, and Medicare and Medicaid providing coverage to others, and still others buying their own health insurance, approximately 95% of the population in Grandview has at least some health insurance coverage.[3]

With such extensive coverage, it might be supposed that people would be happy about the health-care scene in Grandview. In fact they are, but not just because third-party payers help ease the burdens of medical costs. They are quite satisfied with their health-care delivery system as such. Grandview has over 120 physicians, most of them specialists, to serve both the resident population and an equal number of people living in the outlying areas who regularly come to Grandview for medical care. Both groups believe that Grandview's physicians are well-trained, conscientious, and caring.

They are satisfied with the hospital facilities as well. Grandview has three hospitals (Maxi, Meso, and Mini; 400-, 160-, and 130-bed units respectively), and all three are well-equipped and well-staffed. The people are particularly pleased because, as many of them put it, "we have a choice" of where to go for treatment.

The physicians have a similar reason for being pleased. They like the choices they have of where to treat their patients. If they receive less than satisfactory service from one hospital, they can express their disapproval very effectively by admitting patients at one of the other hospitals. Quite understandably, the hospitals do what they can to supply the physicians with good services and high-technology medical equipment. In addition, the physicians are generally happy with their lot since, with the exception of surgeons, there is no oversupply of physicians in Grandview; nor is there an undersupply. So Grandview's physicians make a good living by conscientiously treating their patients, concerning themselves with medical issues as they arise within their own subspeciality, and *with little else.*

This lack of concern for what takes place beyond the physicians's busy schedule points to the two major health-care distribution problems in Grandview. The first is the quality of health care. It is true that the quality of this care is generally high, so that the problem here is not one of disasterously serious proportions. Nonetheless, the quality of health care could be significantly higher if triplication of services were not so common. Each hospital in Grandview, for instance, has a fully-staffed, round-the-clock emergency room. Each hospital also provides obstetrical, pediatric, (overlapping) surgical and other services to the community. The result is that the triplicated service areas are under-utilized in one, two, and sometimes all three hospitals. The further result is that the hospitals with the underutilized service areas have not developed enough staff expertise to provide the highest possible quality service.

The physicians are not wholly to blame for this situation even though their lack of community-wide concern when it comes to health care, and their demands on the hospitals to provide various services and equipment, encourage triplication. Nor are the people living in and around Grandview wholly to blame either, although they too, with their strong desire for a choice of hospitals, encourage triplication as well. Both the physicians and the people must share the blame with the hospitals themselves. All three hospitals operate in a highly competitive setting with each trying not only to do a better job, but to do more jobs. When one hospital offers same day surgery, the other feels that it too must offer this service; and when one hospital purchases a piece of equipment, it becomes almost mandatory for the other hospital to do the same. Again it matters little whether the service or equipment will be underutilized and, therefore, not used as well as it might be. What matters is to stay competitive.

The second problem on the health-care scene in Grandview is cost. The inefficiencies of triplication not only keep the quality of health care down; they also help drive costs up. These rising medical costs are fueled, of course, not just by triplication. A related problem in Grandview, surplus beds (approximately 130 of them), has the same effect. In addition, costs in Grandview are rising because of the inflation pressures found in the larger society, the attempts by its hospitals to play catch-up in remunerating their employees, the purchase of technologically advanced equipment, the defensive medicine practiced by its physicians, and, perhaps most importantly, the lack of concern about costs by either the providers or those served (because direct costs of medical care are borne by third parties: viz., the Federal Government, private insurance companies, and employers).

In order to deal with these twin problems of less-than-optimal quality and high costs, the three hospitals formed a special committee, which they called the Tri-Hospital Committee (THC). The THC was composed of four representatives from each hospital (including the chief administrator, an "affiliated" physician, and two board members) as well as two at-large representatives from the physician community. It was this group that commissioned the 1978 survey and in 1979 retained me as a consultant to see whether something could be done about the triplication and the surplus beds.

The immediate aim of my consulting efforts was to develop a community-wide plan to deal with these problems. Toward this end, I met privately with several administrators at each hospital, with many physicians, as well as with key community leaders. Over the course of several months of negotiating, the following plan emerged. First, the surplus bed problem would be tackled initially by closing between 20 and 30 beds at Mini hospital. The rationale behind this recommendation was that Mini had the lowest daily patient census and therefore would suffer the least by making this sacrifice. Second, Mini would give up its obstetrical and a large part of its pediatric services, so that now these services would be provided mainly at Meso (a Protestant hospital) and Maxi (a Catholic hospital). Again the reasoning here was that since both obstetrics and pediatrics were being grossly underutilized at Mini, it made sense to have this hospital make this sacrifice as well. Third, in return for its sacrifices, Mini was to receive the new equipment projected for ophthalmology and would become the center for the community's ophthalmology surgery. Fourth, Mini would also receive End-Stage Renal Disorder (ESRD) facilities that were scheduled to be introduced into Grandview in late 1980 or early 1981.

There were two supplemental parts to the plan not having to do with consolidation of services as such. The first concerned the status of the Tri-Hospital Committee's directorship. Each of the hospital's chief administrators directed the THC on a rotating basis for four months. This rotating procedure was certainly a fair way of doing things, but it tended to weaken the Committee. It not only did it not allow the Committee to develop any overall sense of direction; it also usually resulted in the director's giving the Committee his undivided attention only after his primary duties at his hospital had been attended to. This meant, in fact, that he would spend a significant amount of time attending to the Committee's business only a day or so before its full-dress monthly meeting. Given these facts of administrative life, the plan's recommendation was that a permanent full-time (or, as a start, a part-time) director be appointed. As it was envisioned, this director could be a retired physician, an "outside" hospital administrator, or some other proven community leader who could act in a neutral (i.e., fair) fashion with respect to the three hospitals and the physician community.

The second supplemental part of the plan was concerned with a kind of understanding about the status of the plan itself. Almost from the start of the negotiations toward developing the community-wide health-care plan, it was obvious that the conditions were not right for doing all that needed to be done. The history of Grandview's health-care delivery system told why. In the early seventies, for example, Maxi hospital invited (some say "bribed") the largest clinic of physicians (about 40 physicians at that time) to build new offices on property adjacent to and owned by that hospital. By accepting this offer, The Clinic changed its pattern of admitting patients away from Mini and toward Maxi. Indeed, this one move assured that Maxi hospital would be the "maxi" hospital in Grandview in the foreseeable future. Not to be undone, Meso hospital got approval for a modernization and expansion program that actually increased its total number of beds at a time when it was already known that Grandview's surplus bed problem was getting serious. Meso also invited a group of "independent" physicians to build offices adjacent to its facilities. There were other examples in recent history of conflict not only among the hospitals, but among the physicians and between the physicians and hospitals. There was, to give one more example, little love lost between many of the physicians in The Clinic and certain independent physicians. Primarily the dispute had to do with competition for patients and the independents' fear that The Clinic was getting too big for the good of the community.

With such high levels of mistrust all around, all that could be hoped for was a modest plan that, if it proved successful, would be supplemented with a second and a third plan. The understanding, then, was that community-wide planning would become an ongoing process. The failure of the first plan to consolidate the emergency rooms from three to two was, therefore, not to be seen as a failure as such. What could not be done during phase one, when the mistrust level was still high, might possibly get done in the future after the community had gained some experience at, and had some success in, community-wide health-care planning.

That was the plan. In fact, this is what happened. Right in the middle of the planning process, Grandview's city government decided that it wanted to get out of the ambulance business, which it had been in for many years, and decided also to invite the hospitals to take over this service. With this invitation, the hospitals had an opportunity either to use the Tri-Hospital Committee in order to incorporate the ambulance service into its community-wide plan, or to have each hospital fight over the spoils. Almost without thinking about its options, it chose the latter course. As it turned out, the main fight developed between proposals presented by Maxi and Meso—the two strongest institutions. However, the dust raised by this conflict was so great that, after the city tentatively accepted Maxi's proposal, it backed off and decided to keep the ambulance service after all.

In spite of the ambulance struggle, the plan seemed to be gaining support from all those who were consulted. Even the ophthalmologists who stood to be inconvenienced to some extent by transferring the majority of their surgery to the more remote Mini Hospital were cooperative. Surprisingly, it seemed that in spite of its history of conflict that Grandview's self-initiated community-planning process was going to succeed. Prospects for the plan improved even more when the appropriate state-governmental agency approved Mini's application for the ESRD facility and The Clinic recruited a physician to administer that facility.

Unfortunately, within the span of a month, the whole planning process completely collapsed. In the end, the ophthalmologists decided to veto their role in the plan because they could not unanimously agree among themselves to move out to Mini. In addition, an alternative proposal for an ESRD facility emerged that undermined that portion of the plan. The alternative proposal for a "free standing" clinic (i.e., one outside a hospital—but, in this case, adjacent to Meso) was presented by an independent physician who had not as yet even moved into Grandview. However, supported by other independent physicians (who, of course, opposed The Clinic's proposal for basing the ESRD facility at Mini), the free-standing proposal gained Federal Government approval and thereby overturned the state government's decision. With nothing to offer Mini in return for its sacrifices (of beds and services), and with a feeling by some members of the THC that they had been double-crossed by other members, the planning process expired.

There is a sequel to this story. Once the hospitals had reduced the Tri-Hospital Committee from a decision-making to simply an information-disseminating body, the attitude of many of Grandview's health-care providers became that of "dog-eat-dog." Indeed, in short order, one of the dogs got eaten up. In spite of its historical antagonisms to its sister Catholic hospital, Mini's board of trustees decided that it would merge with Maxi. During and after the planning process, Mini's daily census figures continued to decline and the board saw that it would survive as an independent hospital only if heroic measures were taken. Upon considering the matter, the board decided that there was no point in continuing the struggle since the intense competition among the hospitals would, it felt, hurt the community more than it would help it.

So with one gesture of resignation on Mini's part, the forces of the market place achieved what the planning process failed to achieve within a span of a year and one-half. It is too early to tell at this writing just what consolidation moves will take place at what can now be called Super-Maxi. It seems probable, however, that obstetrics and pediatrics will be phased out at Mini and that its emergency room will be downgraded to a kind of first-aid station. Also, a certain number of beds will undoubtedly be closed. What Super-Maxi will decide to place in Mini in order to make it a viable facility remains to be seen.

Other changes have taken place since the planning process ceased. For one, the free-standing ESRD facility is in place and functioning well next to Meso. For another, more physicians continue to come to Grandview with the feeling being that their presence is enhancing its medical-care delivery system. Finally, primarily as the result of the merger, the medical community is even more polarized than before. There always was a fear in Grandview that Maxi would swallow up the competition. Since it has done just that, this fear has now been magnified. Not only has this polarization affected emotions within the hospitals, but it has affected some of the physicians as well. Even more than before, they tend to align themselves with either Meso or Super-Maxi. Competition between the two sides (Meso and the Independents, on the one side, and Super-Maxi and The Clinic, on the other) promises to be fierce in the future. Already there is an emerging struggle between Meso and its supporters, who think that a second CT scanner is needed in the community ("After all," as a Meso supporter put it, "what happens when the older Maxi scanner has some downtime—as it is prone to have?"), and Super-Maxi and its supporters, who think that only one new scanner is needed, and that it should be located in the largest hospital facility in Grandview as a replacement for the older model.

Lessons

What, it might be asked, is to be learned from this case study? The first and most obvious lesson is that a United Nations-like organization such as the Tri-Hospital Committee, where each autonomous agent (whether that agent be a hospital, clusters of physicians, individual physicians, or governmental agencies) has a veto over decisions made by it, cannot work. To be effective the THC would have had to have the decision-making powers that absolutely no one was willing to give it. The problem here was not only the level of mistrust, but the medical traditions that give individual physicians God-like powers over where and how they conduct their practice–business, and that encourage hospital administrators to think of hospitals as businesses. With these traditions in place, several of the participants in the planning process were almost incredulous that such a process was even taking place. It was as if they asked themselves: Why should we be asked to plan in this way when other businesses are not?

Assuming for the moment, but only for the moment, that reasons can be given in favor of community-wide planning, Grandview's experience strongly suggests that such planning is not going to be successful if it is locally generated. Some outside power like the Federal Government seems to be needed to give authority to whatever decisions are made. However, if we look to the Federal Government not simply as a power source, but as a power source acting intelligently, Grandview's experience should give us pause. The community-wide planning that took place in Grandview actually was not quite so voluntary as the account of events in the first part of this paper suggests. It is true that the actual planning took place without direct governmental supervision. But the shadows of the Federal Government were everywhere. It was understood by all that federal monies simply would not be forthcoming if planning did not take place both on the state and local levels. Yet although the Federal Government encouraged community-wide planning, it was a federal agency that helped kill the planning process when it rejected the Committee's proposal to place the ESRD unit in Mini Hospital. It was as if the Federal Government were not acting with one mind, to say the least.

It was not acting with one mind in another matter as well. During the time the survey was being conducted, the Federal Government was not only urging Grandview to close some beds in order to deal with the surplus bed problem, it also was urging certain communities surrounding Grandview to close down their rural hospitals to accomplish the same goal. There are several small (30–50 bed) rural hospitals within an hour's drive of Grandview; and, generally, they tend to have low daily census averages. It made some sense, therefore, for Washington to suggest that some of these "inefficient" facilities be closed. However, at the same time, the Federal Government had a policy of encouraging more physicians to move to the rural areas. Again this policy made sense since the physician-to-patient ratios in Grandview and in the rural area that surround that community were 1/900 and 1/3000–4000, respectively. So whereas the Grandview physicians found themselves comfortably busy, the rural physicians were being overwhelmed with work. But as several rural physicians told me in their interview, if their hospitals were closed, they will move into Grandview or some other community where they will have easy access to a hospital. However small their hospitals might be, these physicians felt that they could not serve their communities effectively as family practitioners without them. To these physicians, federal policies that urged hospital closings in the rural areas and physician migration to those areas were simply contradictory.

In this connection, there was evidence that both problems (inefficient operations and physician shortage) were taking care of themselves without federal intervention. Several of these rural hospitals managed to operate with surprising efficiency in spite of low census figures by using part-time help. Many local residents were willing to support their community hospitals by working only when they were needed. As to the physician shortage, certain physicians (especially surgeons), who found that their practice in Grandview was not as large as they had hoped it would be, were beginning to serve the outlying communities on at least a once-a-week basis. So the forces of the market place by themselves seemed to be taking care of these problems.

These forces, again in contrast to governmental ones, obviously worked successfully in Grandview itself. The consolidation of Mini and Maxi into Super-Maxi leaves two healthy hospitals in Grandview with just enough duplication of services to avoid monopolistic inefficiencies, on the one hand, and to avoid triplicative inefficiencies, on the other.

Another market force that helped to improve the health-care system in Grandview is ABC Industries. As has already been noted, this firm's employee benefits package contains generous health-care provisions. However, these provisions were negotiated prior to a time of the rapid rise of medical costs. ABC was therefore surprised and disappointed to see such costs rising almost uncontrollably. Since it could not take away benefits already granted to its employees, it began to champion the Health Maintenance Organization concept of prepayment of medical bills as a way of containing costs. Put roughly, ABC said to Grandview's health-care providers that either someone in that medical community will sponsor an HMO or it will bring an HMO to town itself. Already faced with a surplus bed problem and realizing that an HMO would exacerbate that problem, the hospitals were less than enthusiastic about ABC's proposal. The physicians were also not thrilled by the proposal, since most of them believed that HMO medicine means inferior medicine. Yet with firm but gentle prodding from ABC, The Clinic saw the light and decided that if you cannot beat them, join them. So under the auspices of The Clinic, anyone living in and around Grandview will soon be able to choose between fee-for-service and HMO medicine.

A second lesson to be learned from this case study is that significant progress can be made in providing health-care services with a mixed health-care system.[4] The general argument that supports this claim follows in the next few paragraphs.

One of the underlying assumptions of the mixed system is that the question 'How should health care be distributed?' is not easy to answer. If a person focuses almost exclusively upon the notion of justice in answering this question, the answers usually take an egalitarian form and, these answers do seem easy to come by for such a person. Conversely, if a person focuses almost exclusively upon the notion of freedom in dealing with this distribution question, the answers again seem easy to come by; only now they are in terms of the free marketplace. But Grandview's experiences suggest that these easy ideological answers fail to take into account certain difficulties inherent in both the egalitarian-governmental system and the free market systems. On the egalitarian side, it was obvious that the governmental agencies that would be needed to enforce an equitable distribution of health care did not acquit themselves well in the Grandview planning experiment. As was already noted, governmental policies often were at cross-purposes. Moreover, these policies often seemed far removed from the realities of the local scene,[5] arbitrary, and expensive to implement.[6] Certainly there is nothing original in these charges against the governmental-egalitarian system. Still, even if not original, they were confirmed by what happened in Grandview; and, therefore, they provide evidence against that system.

On the other side, the weaknesses were equally obvious. The issue of high medical costs was never seriously addressed by the hospitals and physicians in Grandview when they were left to their own devices. Their failure to address this issue may have occurred because they are not actually operating in a purely free market environment. It could be argued, after all, that most medical bills are paid by a third party (thereby giving the consumer little incentive to shop around) and most physicians are paid on a fee-for-service basis (thereby giving them little incentive to reduce services or expenditures).[7] Be that as it may, communities like Grandview represent the free marketplace as we know it today, and in that market place the hospitals and physicians were relatively oblivious to rising costs.

In a different sense they were oblivious to, or at least unable to deal with, another problem: viz., the needs of the poor. As we have noted, Grandview is neither an overly wealthy nor overly deprived community. Yet, even when unemployment is not overly serious there, it is clear that the physicians and the hospitals are unable to take on the extra burden of the medical costs that the unemployed normally incur in the course of time. To be sure, some costs can be absorbed. But the majority of the costs are such that without money of their own, the unemployed either could not avail themselves of medical care or would have to be given financial help (more than likely) from the government to obtain such care.[8] It is interesting to note at this point that even some of the most ardent free market physicians I interviewed realized that without Medicare and Medicaid, the level of health in Grandview would have been lower and the size of the medical community (i.e., the number of physicians, etc.) would have been smaller. So even these physicians were willing to grant (in moments of weakness) that the government is needed to help care for the poor, the aged, and those seriously ill who cannot care for themselves.

The weaknesses of each of these positions points to the strength of the other. The governmental-egalitarian system has the resources, and is the only one that has to care for those who cannot care for themselves. It also has the power to control costs, to some extent at least, since it can withhold large amounts of monies from those who do not practice reasonable cost-containment policies. Yet the hospitals operating within a limited free market system did, in fact, accomplish much in consolidating Maxi and Mini—and did so quite efficiently without getting involved in the time and money-consuming process of planning. In addition, all during the planning and post-planning process, Grandview's physicians were continuously working within their subspecialties on procedures that would lessen a patient's hospital stay; or possibly even move certain procedures out of the hospital completely.[9] Even ABC Industries (as a kind of insurance carrier) did its part in the free marketplace toward containing health-care costs. It not only acted as an agent of the market place itself in demanding reasonable medical costs, but it did so by creating new competition for the old fee-for-service mode of payment that is so firmly entrenched in Grandview.

So the mixed system of providing health care, where almost everyone who is heavily involved on the health-care scene either as a provider, payer, or recipient has a say as to how it functions, seems to have some merit. It seems, if Grandview's experience is any indication, to encourage the best within each system and yet to keep the worst under control.

Defense of the Mixed System

In the remainder of this article, I will defend the mixed system against certain objections and, in the process, characterize that system and the two systems that have been traditionally opposed to it, in greater detail. The first objection to the mixed system comes from those who would defend a pure version of the governmental-egalitarian system. They argue that dealing with health-care distribution problems under the mixed system is difficult at best since the responsibility for making decisions is decentralized among such disparate groups as hospital administrators, physicians, insurance administrators, governmental administrators, and consumers. What the governmental-egalitarians prefer instead is some centralized power source that will make planning both consistent and efficient.

Quite apart from the fact that Grandview's experience in planning shows that consistency and efficiency are not necessarily virtues of centralized governmental planning, in part because the Federal Government is itself not a single entity, but a series of conflicting agencies and sub-agencies, this objection to the mixed system is flawed for another reason. The mixed system is designed less to be efficient than to be representative. The theory behind this system is that all those involved in the health-care distribution process should be involved in the decisions about how health care should be distributed. Now it is true that there is some degree of proper representation on the federal (and state) level. The various legislative bodies, executive agencies, commissions, panels, and so on that constitute the power of the Federal Government are made up of people with different backgrounds in the field of providing health-care services. Although some are just plain administrators, others have training and experience in community health, hospital administration, and in medicine itself. Also, of course, some are elected by the people. Nonetheless, as representative as this centralized system is, it cannot be as representative as a mixed system. Like its more centralized counterpart, the mixed system has federal representation, so that the interests of the poor and the people at large can be taken into account. But under the mixed system, the government's share of representation is less than 100% in that the government shares representation with the physicians and hospitals whose interests

and points of view about health care are different from those of the
Federal Government. It also shares representation with such payers as
insurance companies and those institutions that pay premiums to these
insurance companies. Consumers also have representation under this
mixed system by exercising their powers to choose between HMOs
and fee-for-service arrangements, by choosing among physicians, and
even by choosing when to become consumers.

This is not to say that consumers should not become involved in
still other ways if the mixed system is to work as well as it might. During the Grandview planning process consumers were represented only
indirectly through the 1978 survey. Although this survey told the THC
what these consumers thought, once the survey was completed it was
left to the Committee, the physicians and the government to decide
how to reorganize Grandview's health-care delivery system. At that
point, consumers played no direct role in the decision-making process.
Working at its best, then, as a system that features improved representation over the government-run egalitarian system, the consumers of
medical services should be found on all planning committees in "representative" numbers. Moreover, they should be there as consumers
and not, for example, as members of some hospital board.

Ideally at least one other group should be represented in the mixed
system. If delivering health-care services is seen as a complex activity
that gets done best when all points of view about how the job is to be
done are represented, then nurses should also be represented. As was
just noted, although in Grandview the consumers were ignored as far
as the planning process is concerned, they at least were represented in
the 1978 survey. Nurses were not even accorded this honor. For all
practical purposes they were treated as lowly employees rather than as
professionals. If such treatment is typical, and if their roles in delivering health-care services is important enough to affect the quality of
health care delivered to the general population and also to affect the
cost of such care, then they deserve better treatment. With respect to
quality, it could be argued that if the voice of the nursing profession
were more powerful, the current shortage of nurses might not be so
severe as it is today and therefore the quality of health care in the US
might be higher than it is.[10] As to costs, again if the voice of the
nursing profession had been listened to, nurses might be used more frequently in ways (e.g., as physician extenders, in primary health roles
in the rural areas, and so on) that could help cut costs and at the same
time improve the quality of health care.[11]

So the proper response to the first objection that the mixed system
is inefficient because it is decentralized is that: (a) this system may not

be any more inefficient than the remote and complex centralized system; and (b) it is superior to the centralized system in being more representative, where being representative means getting everyone who is touched by medical care services involved in the decision-making process.

A second objection to the mixed system comes from the opposite, that is, the free market, side of the political spectrum. Free market theorists are often apt to argue that their theory can work only in its pure form. For them, the mixed system cannot be expected to succeed even if it contains free market forces because the governmental forces in the mixed system prevent the free market forces from working properly.

Surely one form of this objection has to be wrong, since in Grandview the free market forces effectively solved the problems of consolidation and also helped in a significant way to introduce the HMO concept in that community. All this was done even with a heavy governmental presence on Grandview's health-care scene. If, however, the sense of this objection is that health-care distribution cannot reach its optimal level with respect to quality and quantity of health care unless the free market theory is fully implemented, then it is difficult to know what to say. In effect, what the objection amounts to is that the free market theorists think that they are right and everyone else is wrong. About the only response one can make to such a claim is that even if the free market forces do all or almost all that its admirers claim for them, it is difficult to see how the medical needs of poor people can be met without at least some governmental support. Voluntarism can certainly supplement governmental contributions, but the sums of money needed to provided decent health care to the poor, especially as such services become technologically more exotic, can hardly be dealt with by voluntarism alone.[12]

The final objection to the mixed system comes from the egalitarians, and it is simply that the mixed system is non-egalitarian. In a very real sense this is the most serious objection to the mixed system since it can be construed as saying that unless medical care services are distributed equally, they are being distributed unjustly. It can also be construed as saying that the mixed system is basically not much different from the free market system, in that, like that system, it countenances two forms of inequality. First, it permits physicians and other health-care providers to become wealthy at the expense of the sick; and, second, it allows some people to receive better health-care services than others.

In replying to this objection, it is worth looking at some quantitative data in order to back the claim made earlier that Grandview's residents are satisfied with their health-care delivery system. The respondents in the 1978 survey were asked to comment upon the quality of the health-care system in their community and more specifically, whether, as they see it, it is better or worse as compared to the past. They were also asked comparable questions about the "overall quality of life (in their community)," "the quality of public school education," and "the quality of American automobiles." Their responses are summarized in the table below.

As the data show, comparatively speaking, the health-care system was rated very high. It is true that in response to other questions, there were many complaints about the cost of such services, but no measurable complaints about inequality either with respect to access to such care or the quality of the care itself. As the Grandview residents perceived it, then, the quality of health care in their community appears not only to be high, but to be distributed without any gross inequity. Inequity is not even an issue for them. They realize, of course, that some among them have the financial resources to pay for medical services received in special hospitals and clinics in other communities. Some also realize that certain physicians refuse to take Medicaid patients. Still, even those who complain about these matters feel that the medical treatment they receive is at least adequate, and many say that it is excellent.

To be sure, it could be argued that the residents of Grandview are in no position to assess the quality of the care they are receiving, since they are not physicians themselves and since they have come to a favorable attitude (e.g., because of capitalistic and AMA propaganda) toward the only health-care system that they know much about. Still, it is a bit bizarre to claim that gross inequities are present and yet are not perceived by the vast majority of the population. Prima facie, it would seem that the facts are better explained by saying that a certain amount

Direction of Change in Quality[a]

	Health care, %	Life, %	Education, %	American automobiles, %
Changed for the better	84.0	65.3	47.0	23.8
Changed for the worse	6.5	19.0	30.0	58.5
No change	8.8	14.3	14.8	14.3
No answer	0.7	1.4	8.2	3.4
	100.0	100.0	100.0	100.0

[a] $N = 400$

of inequity is present in Grandview's medical care system, but that the amount is not so great as to be damaging to the health or the spirit of this community's people. The same general argument applies to the part of the egalitarian argument that claims that physicians are overpaid. There is some resentment here; but, by and large, Grandview's residents respect their physicians and see them as being no less deserving of the rewards of their profession than any other group of professionals.

The egalitarian could grant all the facts about how satisfied Grandview's residents are, but argue that these people ought not to be satisfied. He might add that even if the inequities in medical care are not gross (i.e., not so serious as to leave segments of the population uncared for), the inequities in communities like Grandview are still unjust, since they result in some people living longer and better than others.

Addressed to the issue of how health care should be distributed in Grandview, the claim about living longer and better is probably false. Too many of the poor have access to too many (although not all) of the best physicians (and facilities) in the community for this claim to have much substance. Further, as is well known, living longer and better is not simply or even mainly a function of the quality of health care received, but of life style as well. It makes a big difference where one works, and whether one smokes, drinks, and so on. In any case, the claim that health care is maldistributed in Grandview might better be directed at the health-care services of other communities such as those in large urban centers with much unemployment and remote rural communities with no physician within easy reach of people in need of medical care. In other words, the thrust of the claim could be that in illustrating the strength of the mixed system by telling Grandview's story, I have turned my eyes away from the real problems in our health-care delivery system.

At this point it should be recalled that the announced intent of this article is not to prove that the mixed system is the best of the health-care distribution systems available. Rather, in large part, it is to show by illustration that this system can work reasonably well in a community that is typical of many in not being overly endowed with wealth or overly burdened with poverty. Part of the intent, as well, is to suggest how the mixed system should be modified to work better than it did in Grandview. Two suggestions that have already emerged from the discussion are that greater participation in decision-making on the part of at-large citizens and on the part of nurses should be encouraged. An additional suggestion that has not been made as yet, but certainly seems justified by what happened in Grandview, is that the Federal Government should take its cost-cutting regulatory responsibilities less seriously than it has in the past.[13] In the years between 1978 and 1981 it is difficult to point to any regulatory move it made that had any serious effect on containing costs in Grandview. If the government made any contributions toward improving the health care of those living in and around Grandview, it was through Medicare and Medicaid, and in other ways such as providing services and information about community health, sponsoring research, subsidizing the education of health-care providers, and so on.[14] If, then, other communities have problems in distributing health care that are more serious than those found in Grandview, the overall argument of this paper is that the mixed system is one that works reasonably well and is reasonably modifiable to deal with these problems. The mixed character of the mixed system is not jeopardized seriously therefore, if, in some communities where health-care problems are more severe than in others, the Federal Government is the third-party provider for more people than in these other communities. So long as the Federal Government, the insurance companies, the physicians, those administrating hospitals, nurses and the people at large all have a significant say as to how health care is to be distributed, the checks and balances aspect of the mixed system and the system as a whole is safe. The Grandview experience seems to suggest that trouble comes when one group in the mixture takes control of the process of health-care distribution and arranges it to suit its own ideological outlook and its own needs and desires.

Notes and References

[1]This name, the names of the hospitals in Grandview and other proper names used in this article are fictitious.

[2]I carried out both the 1978 survey and the 1979 consulting work as a Research Associate with Jefferson Davis Associates of Cedar Rapids, Iowa and Atlanta, Georgia.

A more complete report of the events described in this article can be found in "A Report from a Community Health-Care Planner," *Health Care Management Review*, (No. 4, Fall, 1981), 49–55 and "A Follow-up Report from a Community Health-Care Planner," *Health Care Management Review* forthcoming.

[3]This figure is probably a little high since it is based on information gathered from telephone interviews. Presumably those without telephones are more likely not to have hospital insurance than those with telephones. Government statistics show that (at about the same time period) almost 80% of the

population is covered by private health insurance while 12% of the population is covered by Medicare: *Statistical Abstracts of the United States, 1980,* US Department of Commerce, Bureau of Census, Washington, DC, pp. 108, 347, 551. Because some people have both Medicare and private insurance, these figures cannot be added together. Also it should be kept in mind that many people without insurance can still receive "free" medical services through Medicaid, and so on.

[4]H. Tristram Engelhardt, Jr., "Health Care Allocations: Responses to the Unjust, the Unfortunate, and the Undesirable," In Earl E. Shelp (ed.), *Justice and Health Care* Dordrecht, Holland: D. Reidel Publishing Co., 1981). On pages 121-122 Professor Englehardt makes these distinctions as follows:

1. The Pure Free Market System

In a pure free market system, one gets what one can pay for and what some-one is willing to sell. The free market system is based on the trading of some goods for health care goods. It turns on the notion that one can exchange the resources one owns for resources that other people own—a notion that works strongly against egalitarian schemes. This has been the prevailing system in North America.

2. Mixed Systems

A partially free market for health care with a decent minimum of health care for all.

In a mixed system (as I mean it here) a decent minimum is provided for all independently of their ability to participate in the market, and in addition, one can purchase additional care if one has the resources, and if someone is willing to sell the services. The second system, which is a two-tiered system, I will refer to as a health insurance scheme, where all are insured against a certain level of costs and are *assured* that they will receive a certain mini-mum amount of treatment. It is obvious that such schemes can range from a Medicaid-Medicare system as we have today . . . to a totally comprehen-sive health insurance scheme, such as a number of proposals that have been introduced in Congress.

3. The Egalitarian System

In a strictly egalitarian system everyone would receive the same standard of care, presumably one that affords at least a decent minimum, and there is no possibility to acquire additional health care on the open market. The third option underlies the notion of an inclusive health service, which attempts to encompass all health care providers and health care recipients. A second sys-tem, a second tier, is, under such a view, held to be improper and is not tolerated. Thus there have been arguments in Britain against allowing a sec-ond free market tier . . . .

[5]Victor R. Fuchs argues against the centralized system. "Physician-centered approaches permit much greater flexibility in adjusting to the needs

and preferences of patients at the local level, where the care is actually provided. Most important, they keep medical decision making where it belongs, in the hands of the physicians." *The New England Journal of Medicine* 304 (No. 24, June 11, 1981), 1487–1490. He also argues that several factors are working to curtail medical expenditures (per physician), and although there will be more physicians around in the next few years, the overall effect will be to put a kind of ceiling on medical expenditures. If this is so, the need for the Federal Government to intervene to contain costs will lessen.

[6]There are here not only the direct costs of governmental regulations, such as paper work, extra meetings needed to interpret regulations, travel expenses to attend hearings, and so on, but indirect expenses incurred because governmental decisions often take months and even years to be made.

[7]Kim Carney, "Cost Containment and Justice," in Earl E. Shelp (ed.), *Justice and Health Care,* Dordrecht, Holland: D. Reidel Publishing Co., (1981), 169–170. Carney lists other reasons as well, such as the inability of consumers to assess the quality of care they are receiving, the practice of third party payers to pay on a reasonable-cost basis (which practice gives physicians and hospitals no incentive to lower costs), the nonprofit status of certain hospitals (in which institutions the profit motive is by definition absent, as are the efficiencies that go with that motive), and the tradition in medicine that prevents physicians and hospitals from advertising their charges.

[8]At present, philanthropy accounts for between one and two percent of medical expenditures (*Statistical Abstracts of the United States, 1980,* US Department of Commerce, Bureau of Census, Washington, DC, p. 106) It is hardly likely that this philanthropy could be expanded to pay the government's share of medical costs which conservatively represents over 40% of the total.

[9]It is interesting to note in this connection the on-going discussions in the *New England Journal of Medicine* that indicate that physicians generally are concerned with procedures that will help contain costs. The following articles and notes appeared in that journal in the span of a few months dealing directly with cost effectiveness: Richard H. Egdahl, "Physicians and the Containment of Health Care Costs," 304 (April 9, 1981), 900–901; Brandon S. Centerwall, "Cost Benefit Analysis and Heart Transplantation," 304 (April 9, 1981), 901–903; Jonathan A. Showstack and Ira D. Glick, "Cost and Efficacy of Ambulatory Versus Inpatient Care," 304 (June 4, 1981), 1431; Russell Hull et al., "Cost Effectiveness of Clinical Diagnosis, Venography, and Noninvasive Testing in Patients with Symptomatic Deep-Vein Thrombosis," 304 (June 25, 1981), 1561–1567; Harry A. Guess, "Bernoulli's Cost-Benefit Analysis of Smallpox Immunization," 305 (August 6, 1981), 341; David A. Haymes, "Physicians and Health-Care Costs," 305 (August 6, 1981), 349.

[10]According to Letitia Cunningham, in "Nursing Shortage? Yes!", *American Journal of Nursing,* 79 (March 1979), 469–480, the nursing shortage is nationwide in both urban and rural areas. What is worse, projections indicate that the demand for nurses will increase in the next few years.

[11]In *Summary of Public Hearings* of the National Commission on Nursing (July 1981, Chicago, pp. 37–43) it was argued that nurses should have more leadership roles than they do now (e.g., have membership on boards of hospitals). The general sense of the report is that nurses do not have the authority to make decisions commensurate with their training.

[12]See footnote #8.

[13]As if to lend support to this point, James Stacey suggests that federal regulations aimed at cutting costs by putting restrictions on purchases of CT scanners have been all wrong since these scanners have proved to be the most useful of the big ticket items to appear on the medical scene in the past 20 years (the other ones being ESRD equipment and treatment and bypass surgery for patients suffering from angina). "Killed by the Cut," *American Medical News,* Oct. 23, 1981, pp. 13–14.

[14]C. Wayne Higgins and John G. Bruhn make similar suggestions in "Health Care Regulations and the Economics of Federal Health Care Policies," *Health Care Management Review* 6 (No. 4, Fall, 1981), 41–47.

Section IV

The Involuntary Commitment and Treatment of Mentally Ill Persons

Introduction

In "Involuntary Commitment and Treatment of Persons Diagnosed as Mentally Ill" Professor Richard Hull presents a systematic and thorough review of the major ethical and legal issues surrounding involuntary institutionalization and treatment of mentally ill persons. Hull begins by delineating four broad issues for consideration: (1) issues involving the concept of mental illness, theories of mental illness, and application of diagnostic terms such as "paranoid"; (2) issues involving the involuntary commitment of individuals diagnosed as mentally ill; (3) issues involving the treatment of involuntarily committed persons; and (4) issues involving the release of persons classified as mentally ill.

Hull considers each of his four broad issues separately. In each case he reviews the current philosophical and legal literature, not only bringing the reader up to date, but also indicating the directions in which additional philosophical work needs to be done. Furthermore, Hull supplements his essay by the addition of an extensive bibliography. Hull's article, together with its bibliography, should be of great value to anyone interested in the legal or moral issues surrounding the involuntary commitment and treatment of mentally ill persons.

In "Mental Illness and Crime" Professor Robert Arrington approaches the topic of the involuntary commitment and treatment of mentally ill persons in a manner quite unlike that adopted by Professor Hull. Rather than dealing with broad, general issues, Arrington critically evaluates one specific proposal for involuntarily institutionalizing mentally ill persons. However, Arrington's purpose is not merely critical, for his criticisms lead him to make certain positive recommendations of his own.

Arrington begins by critically examining the argument of James Humber in "The Involuntary Commitment and Treatment of Mentally Ill Persons." Among other things Humber contends: (1) that mentally ill persons ought to be involuntarily institutionalized only for law-breaking activity, (2) that testimony concerning the mental state of the offender ought not to be permitted during the offender's trial, (3) that lawbreakers ought to be incarcerated for determinate sentences, with sentences fixed according to the seriousness of the crime, and (4) that some mentally ill lawbreakers could be involuntarily treated, provided that certain safeguards are instituted to assure just treatment of the prisoner. In addition, Humber argues that his system would not punish mentally ill lawbreakers, but merely "banish" them to prison for the protection of society.

Though sympathetic to the general thrust of Humber's article, Arrington finds fault with a number of Humber's specific recommendations. First, Arrington contends that what Humber calls banishment is really punishment. Second, he points out that there are perhaps insurmountable difficulties in determining the seriousness of a crime for the purpose of sentencing. Third, he argues that Humber's understanding of the principle of justice is incomplete, and given a proper understanding of that principle it becomes clear that Humber's system does not assure just treatment of all mentally ill lawbreakers. Finally, Arrington argues that if society is to be protected to the fullest it needs to understand the motivations of the lawbreaker.

Given his criticisms of Humber's argument, Arrington proposes that lawbreakers be examined by psychiatrists before or during trial—not for the purpose of determining mental illness, but rather for the purpose of determining *responsiblilty*. As Arrington sees it, it is responsibility rather than mental illness that is the important concept; for some mentally ill lawbreakers might be responsible for their actions, and justice does not require that these persons be treated differently from other wrongdoers. What justice does require is that *nonresponsible* lawbreakers be treated differently from responsible offenders. Finally, Arrington completes his argument by contending that even if psychiatrists do have difficulty in recognizing mental illness, they could determine responsibility for there is a fairly clear set of conditions under which the concept of responsibility is applied.

Involuntary Commitment and Treatment of Persons Diagnosed as Mentally Ill

Richard T. Hull

Introduction

The ethical issues involved in the practices of committing and treating persons diagnosed as mentally ill against their wills are extraordinarily complex, and strike at the very center of Kantian conceptions of personhood.[1] This article seeks to lay out these issues systematically, report on current scientific understanding of, legal precedents for, and the state of philosophical assessment of these practices, and to indicate the directions in which additional philosophical work needs to be done. Much of what I say will apply, with little alteration, to the ethical issues involved in our practices regarding retarded and multiply-handicapped individuals.

A survey of the literature on the ethics of involuntary commitment and treatment suggests that there are four broad groupings of issues:

(1) Issues involving the very concept of mental illness, theories of mental illness, and diagnostic application of terms such as "paranoid," "schizophrenic," "manic-depressive," and the like.

(2) Issues involving the involuntary commitment of individuals to whom such terms have been applied to institutions, hospitals, or other restrictive environments.

(3) Issues involving the treatment of such individuals, through psychotherapy, introduction of pharmacologic agents, surgery on the brain or other body parts, electroconvulsive ther-

apy and other therapies involving subjecting the body to unusual stimuli.

(4) Issues involving the release of persons classified as mentally
ill.

That the first group of issues involves important ethical dimensions
should be evident from the fact that psychiatric labels operate in a variety of ways to classify behavior in contexts where questions of responsibility, competence, and culpability are at issue. That the second
group involves important ethical considerations is evident from the fact
that, given that involuntary commitment entails a restriction on liberty,
questions of whether such restrictions can be adequately justified are at
issue. That the third group involves ethical questions is evident because we know that virtually all of the current and historical treatments
for mental illness involve some potential for harm to the individual,
and that the usual means of defusing such potential for harm of its ethically (and legally) onerous qualities is voluntary consent by a competent individual—putatively absent in cases of involuntary treatment.
That the last group of issues involves ethical concerns is clear because
we see how inevitably the question of protection of both released individuals and others from harm arises out of our scepticism that those
released are cured of the conditions that merited their involuntary commitment and treatment in the first place.

Ethical Issues Involved in Diagnosis of Individuals as Mentally Ill

Thomas Szasz, a psychiatrist, has been the chief critic of the concepts
of mental illness and mental disease. In a series of articles and books,[2]
he has argued the following theses: (1) the concept of mental disease is
a metaphor; (2) when analyzed, applications of the term either apply to
bodily disease "for example, to individuals intoxicated with alcohol or
other drugs, or to elderly people suffering from degenerative disease of
the brain," or to objectionable behavior of persons who "are socially
deviant or inept, or in conflict with individuals, groups, or institutions"; (3) for those in the former group, the ethical principles of ordinary medical decision-making are appropriate; (4) for those in the latter group, the normal social, moral, and criminal sanctions of society
are appropriate; (5) there is no class of individuals whose behaviors are
not properly classified as either normal, consequent upon bodily disease, or properly dealt with through social, moral, or criminal sanc-

tions; (6) therefore, mental illness is an empty category, for the phenomenon does not exist in the field of human behavior. Szasz and others, on the basis of these theses, argue for the abolition of involuntary commitment.

Not surprisingly, there has been a large number of psychiatrists and others who have disagreed with one or more of Szasz's theses. They note that there are broader conceptions of illness than those involving "the demonstration of unequivocal organic pathology." For example, Talcott Parsons, a sociologist, extends the medical model to include "certain forms of social deviance as well as biological disorders," namely, ones characterized "by being negatively valued by society, by 'nonvoluntariness,' thus exempting its exemplars from blame."[3]

However, the more philosophical issues lie in the first two theses. Can it be established by analysis of the concept of mental disease that its applications are either to bodily diseases or to socially objectionable behavior for which criminal sanctions are appropriate? At stake here is an account of the cause or causes of some particular item of behavior. Only by tacitly appealing to some physicalistic account can it seem that "mental disease" involves a metaphor, that there could not literally be a mental cause of some objectionable pattern of (bodily) behavior. Even if one accepts that the thesis of psychophysical correlation is shown to be highly likely by increasingly sophisticated psychophysiological research, it is the research, rather than any sort of conceptual or linguistic analysis, that establishes it.[4]

Moreover, Szasz does not have a category for bizarre behavior that cannot be definitively categorized either as resulting from known organic pathology or as from within the individual's proper domain of responsibility. There is not even the notion of a temporary category for behavior that is involuntary but due to no known physical disorder. It is important to remember that categorization of behavior is not merely an intellectual exercise consequent upon acceptance or rejection of a theory; categorization serves some very important functions of both a pragmatic and procedural character.

One might imagine the system of justice that would be predicated on Szasz's two-fold classification system. If a *prima facie* socially inept, morally otiose, or criminal act could be ascribed to a bodily disease, then a case might well be made for excusing the act from social, moral, and criminal sanctions; but if no such known bodily disease is implicated, the only recourse would be the route of sanctions. Were our science of human behavior complete, our diagnostic powers efficient, and our attitudes enlightened, Szasz's proposals might prove to be the truth of the matter. Until so, though, they strike one as theoretically speculative and as mistakenly taking empirical issues to be conceptual ones.

Having said that, one should add that much of Szasz's criticism of the system of involuntary institutionalization is fair and points to grave injustices that merit swift and drastic redress. One of the most objectionable practices in his view has been that of involuntary commitment of those diagnosed as mentally ill.

Ethical Issues in Involuntary Commitment

Involuntary commitment essentially involves the unwilling loss of liberty, as well as the contingent loss of many other rights. Since liberty is widely regarded as a human right that underlies the political, social, and moral orders, the first task is to inquire into the possible justifications for depriving individuals of liberty on the grounds that they are diagnosed as mentally ill.

Six such grounds have been adduced in justifying civil commitment: (A) the need for protection of property; (B) the need for custodial care (protection from the consequences of unsupervised contact with the natural or social environment); (C) the need of family, neighbors, or general society for relief from the burden of care or contact with the mentally ill person; (D) the need for treatment; (E) the need for protection from self-inflicted harm; (F) the need for protection of others from harm. The second, fourth, and fifth grounds are sometimes subsumed under a common rubric of dangerousness, but since the possible grounds of dangerousness to self and dangerousness to others are not congruent, it is advisable to address these separately. Each of these grounds is discussed *seriatim* in the following sections.

Protection of Property

Recognition of this justification as independently valid for civil commitment elevates protection of property to a position higher than that of protection of liberty. When this is a factor in a psychiatrist's recommendation of commitment, it is usually combined with some other factor such as protection of persons, as in commitment of children for pathologic incendiarism. The lack of normal criminal punishment options in the case of minors may well be a factor in the relatively high incidence of property-related commitment of adolescents. Walker[5] sensibly suggests that simple property offenses be excluded from consideration as grounds for commitment, arguing that (a) measures such

as detention should be used only to "prevent serious and lasting hard-
ship to other individuals, of a kind, which, once caused, cannot be
remedied"; and that (b) most loss or damage to property can be
remedied by recompensation.

Custodial Care and Protection from Harm

Although a mentally ill person's behavior may pose no threat to others,
the loss of touch with reality may be sufficient to render the afflicted
individual unsafe in the normal environmental setting. With this rationale, involuntary commitment is seen as justified under the principle
of beneficence as preventing harm to oneself. However, this involves a
fundamentally different subordinate principle, that of paternalism.
"According to this position, paternalism could be justified only if the
evils prevented from occurring to the person are greater than the evils
caused by interference with his liberty and only if it is universally
justified under relevantly similar circumstances to treat persons in this
way."[6]

Although it is obvious that an alternative to involuntary civil commitment of individuals who are vulnerable prey for others is restriction
of those who would prey on them, not all "environmental" risk dwells
in the exploiters of the weak; mental illness may manifest itself in a
lack of ordinary caution with respect to such daily hazards as traffic
and household dangers, and inattention to normal dietary and hygienic
needs or to special pharmacological regimens. Protective custodial
care is often seen as the only effective measure to preserve the health
and welfare of the mentally ill person. It still might be argued that such
custodial care can be provided in a manner consistent with the principle of liberty, through the ministrations of family or friends. Such support systems are not always available, however; in addition, there is a
question about the obligatoriness of serving in such a role—one that
entails such a considerable burden and compromise of lifestyle as to
constitute an unjust burden that other individuals cannot fairly be required to undertake. (Important societal differences are involved here;
this option of custodial care within the family may well appear to be a
more reasonable one in a social setting where there is a strong tradition
of multigenerational families with various members not pursuing independent careers outside the home available to fill the supervisory role.)

Hence, the protection of an individual from the distortions of his
or her own mental disorders through civil commitment comes to be regarded as legitimate by virtue of a complex application of both the
principle of beneficence and the harm principle, preventing harm to the
individual (beneficence) and to others who would be unduly burdened
by the duties of care (harm).

Relief from the Burden of Care

In Wyatt v. Aderholt,[7] Governor George Wallace argued that "the principal justification for commitment lies in the inability of the mentally ill and mentally retarded to care for themselves. The essence of this argument is that the primary function of civil commitment is to relieve the burden imposed upon the families and friends of the mentally disabled. The families and friends of the disabled, the Governor asserts, are the 'true clients' of the institutionalization system."[7] Wallace concluded that "(T)he providing of custodial care alone is a tremendously important consideration to patients, their families, and the public-at-large." Presumably, the appeal here is to the harm principle; it is difficult to understand what the moral force of relief of the burden of care would be if that were not the implicit rationale. But such an implicit justification raises the question of what degree of harm is necessary to offset the loss of liberty suffered by the committed individual. It also raises the question of whether the psychiatrist has his or her chief obligation to serve the interests of the family and the state, or to serve those of the patient.[8]

Governor Wallace's arguments were rejected; and in the series of cases beginning with Donaldson v. O'Connor and ending with Wyatt v. Aderholt, federal courts have held that (1) nothing justifies the state in involuntarily hospitalizing a mentally ill person through civil commitment procedures except need for treatment or protection of self or others from a clear danger posed by the individual; (2) for an individual who has been subjected to civil commitment, the 14th Amendment's due process clause provides a constitutional right to treatment for the mental disorder that offers the chance for eventual restoration of liberty; (3) only if the disorder is such (as in severe mental retardation or chronic, unremitting psychosis) that treatment would be inappropriate because ineffectual, can mere custodial care be provided, and that only if certain standards of care are met. The ability of psychiatrists or other social scientists to predict dangerousness to self or others thus becomes increasingly important to the commitment process as the courts have limited the grounds on which commitment may be predicated and underscored the potential for serious abridgment of constitutional rights inherent in such proceedings.

Need for Treatment

The need for treatment is a tempting ground for involuntary hospitalization. Under the supposition that various behaviors or states of individuals are attributable to disease (whether physiological or mental), the protective, efficient, equipped wards of the psychiatric institution appear to many to constitute the only appropriate site for treatment. Further, those perceived to be in need of such treatment may well deny their need, either as a further delusional product of their disorders, or in the (perhaps legitimate) belief that their needs would not be well served in an institutional setting.

Generally, the courts have swung about on this as a ground for commitment. The current trend seems to be that the rights of privacy and self-determination from which arise the right to refuse medical treatment also yield a right to refuse treatment in a psychiatric institution in all but critical, emergency situations. Certain treatments cannot ever be given in some jurisdictions, without consent, because of their risks or aversive character. This, together with the nondangerous patient's right to refuse any treatment, and the historical dearth of adequate treatment facilities, effectively undercuts the need-for-treatment rationale. As Judge Bazelon observed,[9] "Absent treatment the hospital is 'transform(ed) . . . into a penitentiary where one could be held indefinitely. . . .'"

These issues also relate to insanity defenses. "Conceptually an acquittal by reason of insanity should lead to release, and if deprivation of liberty can be justified by all, it can only rest on a need for treatment. . . ."[10] If the right to treatment is granted or activated only for those who wish to exercise it, then there is no faulting the logic of one who, acquitted on an insanity plea and committed for treatment purposes, refuses treatment and demands release. On the other hand, at least one commentator suggests that there may well devolve on such an individual a duty to be treated, such that it is clear that involved in a successful insanity plea is an obligation to accept whatever treatment is currently available.[10] It is unclear whether this suggestion also entails a duty to be cured as a condition of restoration of liberty, and a duty to remain confined until a treatment can be developed in case one is not currently available. It may well be that, absent treatment, it is better to go with Szasz's suggestions (also, cf Humber[11]) and turn to the criminal process; some courts have elected this rationale.[12]

Dangerousness to Self

Feinberg[34] has distinguished between strong and weak paternalism: the former involves liberty-limiting interventions in genuinely rational, relevantly informed actions that would tend to result in physical harm to the agent; the latter involves restrictions imposed in the face of evidence that the agent's actions are not voluntary, are not relevantly informed, are in the grip of unreasonable fears, or are being influenced by toxic substances or by severe depression. The principle of liberty conflicts with the harm principle here in that, if we follow Mill (and, arguably, Kant) in allowing unrestricted self-regarding autonomous behavior irrespective of personal consequences, our only justification for intervening will consist either in our need to determine whether an irreversible decision is genuinely autonomous, or in our well-founded belief that it is not.

A decision that is autonomous, in the sense of being free from coercion, knowledgeable of alternatives and of relevant consequences, and accepting of potential risks, is one in which possible negative consequences to the agent do not acquire the character of harms, but rather that of losses. As such, it is plausible to exempt such anticipated, autonomously accepted consequences from the harm principle and to regard as morally indefensible paternalistic interventions in such cases. That is, since the harm principle justifies restriction of liberty only when doing so prevents harm, if the only negative consequences that can be foreseen accrue just to the agent, and the agent has accepted their possibility, those consequences would not be harms whose possibility could serve as the basis for invoking the harm principle. Just as knowledgeable consent is the chief difference between conscription and legitimate military service, charity and theft, sexual relations and rape, so it is the difference between harm and loss.

However, such an application of the principle of autonomy has the character of a limiting case. At the other extreme falls behavior with negative consequences of individuals who wholly lack the capacity to consent. In these cases, the harm principle justifies (indeed, requires) limitation of liberty on behalf of the welfare of the infant, the profoundly and severely retarded, and so on. The ethically troublesome cases lie between; and, as argued by Wear,[24] competence and autonomy are not all-or-none capacities, but manifest degrees and ranges. Hence, the harm principle may justify some restriction of liberty that falls short of that appropriate for absolutely nonautonomous individuals, such as required periodic attendance at an out-patient facility.

Other subtleties compound these issues. One of the most difficult is the iatrogenic character of the loss of autonomy experienced by individuals who undergo institutionalization. That is, commitment, whether voluntary or involuntary, often in and of itself produces or exacerbates incompetence and diminishment of autonomy, increases bizzare and potentially harmful behavior, and the like. Often, such perceptions appear to be a function of the perceptual set of the staff of the institution; studies abound showing that frequently the perceptions of inmates about the mental condition of a person shamming psychotic symptoms are more accurate than those of staff.[13] Another is that the potential for harm to self may be a function of one's external situation,[14] rather than to an internal disorder of the psyche.

But the chief ethical concern over the dangerousness ground for commitment lies with the ability of psychiatrists and social scientists to predict dangerousness with sufficient reliability both to reduce instances of harm to self (or others) and to minimize the number of false positives—persons identified as at risk who, if left unconfined, would not perpetrate harmful acts. As the need for treatment declines as a rationale for non-emergency confinement, one can note two trends in involuntary commitment: the limitation of such hospitalization for periods greater than a couple of weeks "to persons who present an imminent threat of taking their own lives or an imminent threat of substantial physical harm to others"; and limiting the criteria for involuntary commitment to dangerousness to self or others.[15] One effort at studying patients hospitalized following suicide attempts in order to devise predictive devices to identify impending suicides produced measures that would have yielded over one-half false positives; another, which identified a high-risk recidivist group in a 10-year followup study of attempted suicides, gave 67% false positives for subsequent successful suicides, and 46% false positives for subsequent suicide attempts (whether successful or not). Other predictive devices appear both to miss a substantial portion of suicides and to involve some significant number of false positives; the studies were additionally flawed in that they involved only voluntarily committed patients.

Data on pure clinical judgment accuracy have not been gathered and studied in much detail, but the few studies that have included such data indicate varying degrees of accuracy, often reflecting differences between impressions gained through extensive contact with a patient (relatively more accurate) and those gained in the brief contacts preceding a commitment decision. Hence, commitment for evaluative, predictive efforts may require extensive revision of our sense of what is just in the pursuit of data pertaining to long-term commitment justifications.

Dangerousness to Others

With respect to the question of how dangerousness to others justifies deprivation of liberty, the basic arguments turn on what is called the harm principle. Though John Stuart Mill has given the principle its classic articulation, the following statement of it will suffice for our purposes: "It is morally justified to prevent harm to (other) persons when the harm is caused or would be caused by those whose liberty is

restricted.''[6] This principle, in turn, rests on the principle of beneficence, understood as a duty to produce good, prevent harm, and remove harm.[6]

In light of the phrasing of the harm principle (''when the harm is
caused or would be caused''), this issue may seem to be divisible into
the question of justifying commitment in light of harmful acts already
committed, and the question of justifying commitment in order to prevent harm that would otherwise be expected. However, the two questions reduce to the latter.

Deprivation of liberty after a harmful act has been committed is
generally subsumed under the concept of punishment, and punishment
is appropriate only where guilt is appropriate. The development of the
insanity defense has permitted those for whom that defense is successfully raised to escape criminal commitment for the purpose of punishment by virtue of escaping the finding of guilt. Of course, individuals
who are found innocent by reason of insanity are very frequently subjected to involuntary civil commitment, but the justification is not simply because of the fact that they have committed harmful acts (where
that may well be sufficient for involuntary criminal commitment);
rather, the justification is either to obtain treatment for a continuing
disorder, or to protect others from further harm because of continuing
dangerousness. This view is further reinforced by the occasional case
in which innocence by reason of temporary insanity is successfully
pleaded, followed by a recognition that the individual in question no
longer suffers from the temporary disorder and thus neither constitutes
a continuing danger to others nor stands in need of treatment, and this
results in neither civil nor criminal commitment. [In State of New Jersey vs. Lester Zygmanik, the latter was successfully defended against
the charge of first degree murder on the grounds that sleep deprivation
and other long-standing stresses had placed the defendant in a temporary psychotic state in which he did not know the quality of his act (the
harmfulness of shooting a paralyzed brother while in the intensive care
unit of the hospital) and did not know that what he was doing was
wrong. Zygmanik was found innocent; since there was no evidence of
persistent psychosis, no effort was made to obtain commitment for either treatment or protection of others from future harmful acts.][16]

Having committed a harmful act, together with a diagnosis of
mental illness, is often sufficient for civil commitment under the harm
principle, but it has not historically been a necessary condition as well.
Prospective dangerousness to others is often predicated on evidence
weaker than retrospective dangerousness. The chief source of such predictive evidence is the same as the source of evidence of mental
illness—expert testimony of psychiatrists or psychologists. However,

courts have often accepted testimony of family, friends, and neighbors. Considerable doubt has been cast upon both types of judgment. These doubts concern both substantive and procedural components.

The substantive issues are similar to those involved in predicting dangerousness to self. False positives as high as 72% are reported, although some studies comparing the judgment of the courts with that of clinical staff suggest the relative superiority of the latter.[15] Procedurally, the use of counsel and of the right to jury evaluation, together with cross-examination and other rules of evidence, would go far toward protecting the potential patient. Some would hold that adherence to such standards would result in the release of virtually everyone involuntarily confined on dangerousness grounds, since prediction of dangerousness is not well-validated on any known measure. Others would maintain that such measures are too extreme, involving implicit appeals to standard of proof appropriate only in criminal proceedings. There is a serious question, however, whether anything short of the most stringent procedural standards is appropriate, since in fact such confinements may have the character of imprisonment without realistic hope of release, because of the poor treatment situations still existing in many state hospitals.

The inherent logic of civil commitment in order to prevent harm to others dictates an indefinite period of commitment. In this justification, commitment should last as long as the committed individual constitutes a threat. However, the principle of liberty (viz., an individual has a right to the greatest amount of liberty consistent with an equal amount of liberty for each other individual) has suggested to some that an external limit on the power of civil commitment should be imposed so as to preserve some real content for the principle of liberty as it applies to the committed individual. The effort to balance the considerations of each principle in the commitment situation results in an acknowledgment of a right to such treatment as offers hope for restoration of liberty, and severely limits the contexts in which purely custodial, protective care may be offered to only those individuals in whom the mental illness that makes for dangerousness to others is chronic, or unable to be cured or controlled through any known mode of treatment.

Ethical Issues in Involuntary Treatment of the Mentally Ill

Two themes dominate the issues in this range of our concerns. The first concerns whether, under what conditions, and what sorts of treatment may be administered to an involuntarily committed patient. The second has to do with the conditions under which an involuntarily committed patient may effectively refuse treatment for his or her illness.

It must be borne in mind that the courts have approached this question from the perspective of the preceding section's issues. The right to treatment for psychiatric illness has been asserted as a *quid pro quo* right, acquired in exchange for the right of liberty lost through involuntary commitment. And this right was articulated against the position that the state's only obligation to the incarcerated mental patient was custodial care. Thus, the courts did not address, even in the narrowly proscribed area of mental illness, the question of whether there is a constitutional right to treatment enjoyed by all by virtue of humanity or citizenship. The questions raised in recent years concerning a right to health care that might be brought to bear on treatment of mental illness issues have not surfaced in the various decisions involving involuntarily committed persons.

What has been a matter of concern is whether the involuntary commitment of a mentally ill person, together with that person's need for treatment, provides a sufficient justification for compulsory treatment. Chief Justice Burger held, in his concurring opinion to O'Connor v. Donaldson,[17] that committed individuals do not lose their right to refuse treatment, and that there is no basis for compelling treatment since it is the case that the patient's cooperation is essential for most forms of treatment to be effective. One may question whether this latter assertion is correct, particularly since there is an increasing medicalization of our understanding of psychiatric disorders as arising from such factors as neurotransmitter and receptor site disorders[18,19]; his comment seems more appropriate to psychotherapy. A deeper issue, however, concerns the very possibility of ethically treating the mentally ill. If one held the position that involuntary commitment because of a psychiatric disorder leading to dangerousness entailed, or even was presumptive evidence for, incompetence to consent (as argued by the defendants in Rogers v. Orkin,[20]) and if one also holds that the mentally ill patient's refusal of treatment is indefeasible, then only nondissenting mental patients who voluntarily accept treatment could be treated, and then only on the authorization of a proxy or the courts. It would seem that many of those most in need of treatment could not be provided with it. And, such individuals then become effectively imprisoned by their own illness, involuntarily confined and unable to obtain freedom because of a "refusal (that) may be symptomatic of their illness."[21]

The courts have generally avoided this logical trap. The informed consent doctrine has been dissociated from the state's power to involuntarily commit. This is partly because of the tendency in recent decisions to disallow the need for treatment justification as a sufficient basis for involuntary commitment. However, the right to make treatment decisions is not held to be indefeasible. Rather, competence to consent to treatment and/or to refuse treatment is treated as a separately determinable matter, on the grounds that "Mentally ill patients may or may not be competent to make any number of decisions, including the decision to accept or reject medications."[22] Procedurally, courts have held that "(1) an involuntary mental patient may have a right to refuse medication in the absence of an emergency, founded on constitutional right of privacy, and (2) in the absence of an emergency, some due process hearing is required prior to the forced administration of drugs," and the patient has the right to be represented by counsel.[23]

Thus, the courts appear to have endorsed the view[24] that competence should not be viewed as an all-or-none phenomenon, but rather in a domain-by-domain manner (e.g., competence to manage financial affairs, competence to decide on treatment questions, competence to vote, and so on). Although there have been efforts made by both psychiatrists[25] and philosophers[26] to clarify it, we seem short of a well-thought through, objective, operationalized concept of competence that would apply to each of these domains. In its absence, the borderline, "grayer" cases will continue to provoke controversy and dispute within and outside institutional settings.

The courts have rather consistently extended the dangerousness justification for involuntary commitment to involuntary treatment. If treatment is necessary to reduce a patient's level of dangerousness to others or to self,[27] it may be undertaken in an emergency situation where physical harm is imminent and physical restriction or isolation is impossible or ineffectual. However, outside the context of an emergency, the presumption of competence and the right of privacy impose various due process requirements.

Finally, right to treatment issues in the context of the mentally ill have been inconclusively addressed by the courts in several decisions involving *in extremis* treatment decisions. The courts have, in the Quinlan, Saikewicz and Brother Fox decisions, articulated conflicting procedures, at one time leaving prognosis issues up to families and physicians, at another requiring proof in court using a "clear and convincing" standard, in one case disallowing statements made by the patient in healthy contemplation of catastrophic medical possibilities, in another accepting a "living will" procedure. As Annas notes, judges are attempting to legislate termination of treatment issues, and to insert public and judicial review into matters that have been traditionally the purview of the physician, patient, and family.[28] The net effect may well be to encourage physicians not to seek recourse to the courts for tough decisions, but to make them carefully and circumspectly with the active contribution of patient and family[29] when possible. The frequently politically tinged role of the hospital administrator in such decisions as involve involuntarily committed patients *in extremis* may limit the growing aversion to court intervention in that context, however.

Ethical Issues in Release and Resocialization of the Mentally Ill

As indicated at the outset, scepticism exists about the efficacy of much of the so-called treatment for mental illness, as well as its availability to the involuntarily committed patient. One of the more interesting levers that has been used to obtain social reform, first by commentators and later by the courts, is the argument that an involuntarily committed patient has a quid pro quo right to treatment, from which it follows that if such a right is not to be realized, that patient must be released. Both commentators and courts have relied upon social pressure to force legislatures to fund treatment measures, rather than to resort to wholesale release of involuntarily committed individuals.[30,31]

However, uncertainty as to what is required by the law has prompted some to argue for abandonment of various treatment options and modes[27] that may be the only possible options for certain patients. If treatment becomes thus impossible, and need for care and other justifications are lost, then either custodial care or release become the only options. Added to this is the deinstitutionalization movement, found both in the area of mental retardation and chronic mental illness, which provides political pressure on legislatures to close institutional facilities and fund half-way houses and home care as superior alternatives.[32] Further, we have the views of Szasz and others that the notion of mental illness, and thus of treatment of mental illness, involves fictions. On this view, no person may legitimately be involuntarily committed because the constitutional requirement of treatment as a *quid pro quo* right cannot be met. Finally, there are those who are sceptical of the existence, or forseeable likelihood, of effective treatment facilities, and those who believe that individualized treatment plans cannot be effective because of the state's unwillingness to adequately fund requisite staffing; they fear the "danger that after implementation of reform, the same abuses will emerge again in new, though initially disguised, forms."[10]

Such movements and scepticism about treatment realities, and accessibility to effective representation of counsel, together with the poor showing of both clinical impression and predictive measures of psychologists, all contribute to the institution's growing tendency to release individuals who have been involuntarily committed. Add to this the fact that such individuals often return to the environments in which their symptoms of dangerousness were initially elicited (a fact that can only be exacerbated by the decline in public funding of social services), cap it all off with increasing media coverage of instances of violent recidivism, and one may well anticipate a great public outcry against the liberalization of our practices towards the mentally ill. "(L)argely unconscious feelings of apprehension, awe and anger toward the 'sick,' particularly if associated with 'criminality' . . . must be recognized {in our} enormous ambivalence toward the 'sick' reflected in conflicting wishes to exculpate and to blame; to sanction and not to sanction; to degrade and to elevate; to stigmatize and not to stigmatize; to care and to reject; to treat and to mistreat; to protect and to destroy."[33] That ambivalence, together with the uncertainties of prediction, treatment, and indeed the very conceptions of mental disease, will continue to occupy law, social policy, and philosophical reflection for some time to come.

Notes and References

[1]J. G. Murphy, "Incompetence and Paternalism," *Archiv für Rechts und Sozialphilosophie* 60 (1974), pp. 465–486.

[2]T. S. Szasz, "Involuntary Commitment: a Form of Slavery," *The Humanist* 31:4 (July/August, 1971), pp. 11–14; *Law, Liberty and Psychiatry* (New York: Macmillan Co., 1963); *The Myth of Mental Illness* (New York: Harper and Row, 1961).

[3]P. Chodoff, "The Case for Involuntary Hospitalization of the Mentally Ill," *American Journal of Psychiatry* 133:5 (May, 1976), pp. 496–501.

[4]R. T. Hull, "On Getting 'Genetic' Out of 'Genetic Disease'," in J. W. Davis, B. Hoffmaster, and S. Shorten (eds.), *Contemporary Issues in Biomedical Ethics* (Clifton, NJ: Humana, 1978), pp. 71–87; "On Taking Causal Criteria to Be Ontologically Significant," *Behaviorism* 1:2 (Summer, 1973), pp. 65–76.

[5]N. Walker, "Dangerous People," *International Journal of Law and Psychiatry* 1 (1978), pp. 37–50.

[6]T. Beauchamp and J. Childress, *Principles of Biomedical Ethics* (New York: Oxford, 1980).

[7]Wyatt v. Aderholt (503 F 2d 1305), (1974); Wyatt v. Stickney (344 F Supp 373), (1972).

[8]M. A. Peszke, "Duty to the Patient or Society: Reflections on the Psychiatrist's Dilemma," in S. F. Spicker, J. M. Healey, Jr., and H. T. Engelhardt, Jr. (eds.), *The Law-Medicine Relation: a Philosophical Exploration* (Dordrecht: D. Reidel, 1981), pp. 177–186; M. A. Peszke, G. G. Affleck, and R. M. Wintrob, "Perceived Statutory Applicability Versus Clinical Desirability of Emergency Involuntary Hospitalization," *American Journal of Psychiatry* 137:4 (April, 1980), pp. 476–480.

[9]Rouse v. Cameron (373 F 2d 451), (DC Cir 1966).

[10]J. Katz, "The Right to Treatment—an Enchanting Legal Fiction?" *University of Chicago Law Review* 36 (1969) pp. 755–783.

[11]J. M. Humber, "The Involuntary Commitment and Treatment of Mentally Ill Persons," *Social Science and Medicine* 15F:4 (December, 1981), pp. 143–150.

[12]S. M. Goodman, "Right to Treatment: the Responsibility of the Courts," *Georgetown Law Journal* 57 (1969), pp. 680–701.

[13]D. L. Rosenhan, "On Being Sane in Insane Places," *Science* 179 (January 19, 1973), pp. 250–258.

[14]H. A. Prins, "Dangerous People or Dangerous Situations? Some Implications for Assessment and Management," *Medical Science Law* 21:2 (1981), pp. 125–133.

[15]G. E. Dix, "'Civil' Commitment of the Mentally Ill and the Need for Data on the Prediction of Dangerousness," *American Behavioral Scientist* 19:3 (January/February, 1976), pp. 318–344.

[16]P. Mitchell, *Act of Love* (New York: Knopf, 1976).

[17]O'Connor v. Donaldson (422 US 563), (1975).

[18]D. X. Freedman, *Biology of the Major Psychoses: a Comparative Analysis* (New York: Raven Press, 1975).

[19]E. Usdin, D. A. Hamburg, and J. D. Barchas (eds.), *Neuro-regulators and Psychiatric Disorders* (New York: Oxford, 1977).

[20]Rogers v. Orkin (478 F Supp 1342), (USDC Mass, 1979).

[21]Editors, "The Supreme Court Sidesteps the Right to Treatment Question: O'Connor v. Donaldson," *University of Colorado Law Review* 47 (1976), pp. 299–323.

[22]G. J. Annas, "Law and the Life Sciences: O'Connor v. Donaldson, Insanity Inside Out," *Hastings Center Report* 6:4 (August, 1976), pp. 11–13.

[23]H. Creighton, "Rights of Mental Patients," *Supervisor Nurse* 12:5 (May, 1981), pp. 16–17.

[24]S. Wear, "Mental Illness and Moral Status," *Journal of Medicine and Philosophy* 5 (1980), pp. 292–312; "The Diminished Moral Status of the Mentally Ill," in B. A. Brody and H. T. Engelhardt, Jr. (eds.), *Mental Illness: Law and Public Policy* (Dordrecht: Reidel, 1980), pp. 221–230.

[25]L. H. Roth, A. Meisel, and C. W. Lidz, "Tests of Competency to Consent to Treatment," *American Journal of Psychiatry,* 134:3 (March, 1977), pp. 279–284.

[26]J. G. Murphy, "Incompetence and Paternalism," *Archiv für Rechts und Socialphilosophie* 60 (1974), pp. 465–486.

[27]J. W. Cook, K. Altman, and S. Haavik, "Consent for Aversive Treatment: a Model Form," *Mental Retardation* (February, 1978), pp. 47–51.

[28]G. J. Annas, "Quinlan, Sakowicz and Now Brother Fox," *Hastings Center Report*, 10:3 (June, 1980), pp. 20–21.

[29]M. Siegler, "Critical Illness: the Limits of Autonomy," *Hastings Center Report* 7:5 (1977) pp. 12–15.

[30]M. Birnbaum, "The Right to Treatment," *American Bar Association Journal* 46 (May, 1960), pp. 499–505.

[31]D. L. Bazelon, "Implementing the Right to Treatment," *University of Chicago Law Review* 36 (1969), pp. 724–754.

[32]D. B. Wexler, "Mental Health Law and the Movement Toward Voluntary Treatment," *California Law Review* 62:3 (May, 1974) pp. 671–692.

[33]J. Katz and J. Goldstein, "Abolish the Insanity Defense–Why Not?" *Journal of Nervous and Mental Disease* 138:57 (1964), pp. 65ff.

[34]J. Feinberg, "Legal Paternalism," *Canadian Journal of Philosophy* (1971), pp. 105–124.

Mental Illness and Crime

Robert L. Arrington

Philosophers and members of the academic community in general often are accused of treating social issues on such a refined level of abstraction as to make their contributions to the resolution of these issues insignificant. Indeed, they frequently are said to misunderstand social problems, lacking as they do any first-hand experience with the concrete details and dilemmas that confound the politician, the judge, and the policeman. These charges are especially clamorous with respect to crime and society's response to it. Philosophers have developed abstract theories of punishment that attempt to show under what conditions punishment is justified, but seldom if ever do they confront the agonizing perplexities involved in arrest and sentencing that those in decision-making positions must face daily. To many observers, the gap between philosophical theory and social fact is so great as to invalidate the former. Recently, however, philosophers have been attending more to applied issues, and some of them are acknowledging the practical difficulties involved and proposing solutions that take these difficulties into account. James Humber is such a philosopher, and his recent article "The Involuntary Commitment and Treatment of Mentally Ill Persons"[1] contains his practical proposal for resolving some of the dilemmas we confront regarding crime, psychiatric testimony, and the mentally ill lawbreaker.

The problem of the relationship between crime and insanity has concerned thoughtful persons for centuries.[2] What should we do with a lawbreaker who appears to be mentally incompetent? Should he or she be found guilty, or should he or she be acquitted on grounds of insanity or diminished responsibility? In either event, what should be done with these lawbreakers? These questions raise a host of other issues. How are we to determine that a person is mentally ill, and that the illness led to the person's crime? Should psychiatric testimony be allowed in the

149

courtroom, and if so, when and under what conditions? For years psychiatrists were critical of the way courts dealt with the mentally incompetent, arguing that the law was insensitive to the psychiatric conditions and backgrounds of criminals and maintaining that the limited psychiatric testimony permitted in the courtroom was prejudiced and distorted by inadequate, confused models of insanity. Gradually, the psychiatric community began to gain ground in its efforts to reverse this situation. New conceptions of mental illness were introduced into the law, and psychiatrists came to play a more pivotal and independent role in assessing the state of mind of accused persons and in determining what should be done with them.

Unfortunately, this change with respect to the kind and amount of psychiatric input into criminal trials and sentencing has not brought about an obvious, unchallenged step toward a more rational treatment of crime and insanity. Psychiatric testimony is still largely confined to the adversarial procedures of courtroom process, and skeptics note that the nature of a psychiatrist's testimony seems to depend on which side he or she represents, that of the prosecution or that of the defense. At the same time, more theoretical problems are beginning to surface. It has become evident that psychiatrists themselves disagree over general principles of psychiatry as well as over individual diagnoses and prognoses. In recent years our society has become increasingly wary of the entire psychiatric dimension of the criminal law. A backlash has developed, and a growing number of voices have complained that the advent of psychiatric involvement in the criminal courts has led to nothing more than a mockery of justice, a "soft" attitude toward criminals, and a threat to the protection of society.

It is in this context that Humber responds to the problem of how society should deal with offenders who appear to be mentally ill. He too is skeptical of the accuracy and objectivity of psychiatric testimony. Scattered throughout his essay are at least six reasons that might be given for calling this testimony into question:

(1) Psychiatrists do not agree whether mental illness is a myth or not.
(2) Psychiatrists do not agree on a definition of mental illness.
(3) Psychiatrists do not agree on how to identify or classify specific disorders.
(4) Psychiatrists have no understanding of what constitutes adequate treatment.
(5) Psychiatrists cannot show that hospitalization and treatment are in the best interest of certain people in the sense that such care would actually help these people more than no treatment at all.

(6) Psychiatrists cannot predict which patients are dangerous to
themselves or others.

Humber himself does not think that mental illness is a myth, but he
does seem to accept the other propositions. In fact, it is his rejection of
(1) and acceptance of (2)–(6) that raises the following dilemma for him
(and for many of the rest of us): given that mental illness is a reality,
but a reality only dimly understood at the present time by psychiatry,
what role should psychiatric testimony and treatment have in criminal
law and procedure? The problem is exacerbated by the legal and moral
abuses that may result from an uncritical reliance on psychiatric judgment and treatment. Horror stories abound about how persons have
been deprived of their liberty by involuntary hospitalization and kept in
a state of subjection indefinitely even though they never were insane or
overcame their problems long before they were released. Given the extremely limited state of psychiatric knowledge, involuntary commitment and indeterminate confinement are fraught with legal and ethical
dangers.

To deal with the above difficulties, Humber proposes a complex
plan for incorporating psychiatric treatment into criminal law and procedure. First I shall describe this plan in its several steps, and then I
shall give his justifications for these steps. Finally, I shall criticize certain aspects of his proposal and make recommendations for modifica-
tions in it.

Until a person has actually committed a crime, that person should
not, according to Humber, be detained, hospitalized, or incarcerated
on the grounds that he or she is a threat to society or to him- or herself.
If a person P is alleged to have committed a criminal act, then, regardless of P's apparent state of mind, P should be brought to trial to determine whether indeed he or she committed the act. If it is so determined, P should be found guilty, again without regard to P's present
state of mind or state of mind at the time the act was committed. There
is to be no opportunity for psychiatric testimony during the trial. In the
sentencing stage, P is to be sentenced in accordance with a system of
exclusions or banishments that correlates length of incarceration with
the seriousness of the crime, as determined by the political process. No
matter what the crime, the sentence must be determinate, and P is required to serve its full term. After sentencing, P would be examined by
a group of psychiatrists of diverse theoretical approaches, who would
attempt to ascertain whether P is sane or mentally ill, and competent or
incompetent. If *all* the psychiatrists agree that the prisoner is mentally
ill and incompetent, then treatment may be imposed; if found competent, treatment may not be involuntarily imposed, but the prisoner may
voluntarily seek it. Treatment may continue until such time as P's sen-

tence ends. *P* may not be released early because *P* is considered cured, nor indefinitely retained until he or she is judged to be cured. When the sentence expires, *P* must be freed, ill or not, competent or not.

It is Humber's skepticism regarding psychiatry and his concern about possible abuses that lead him to disallow preventive detention and involuntary hospitalization. Psychiatrists, he thinks, cannot determine who are and who are not threats to society and/or themselves. Moreover, they cannot assure us that involuntary hospitalization is in the interest of their patients. Even if there is a right to adequate treatment, psychiatrists cannot claim to know how to provide such treatment. Humber's skepticism also leads him to exclude during a trial any psychiatric inquiry into the accused's state of mind. In this respect he is close to Lady Wootton, who thought it was impossible to determine whether the accused was responsible or not for his or her actions.[3] Although he does not speak of it in these terms, Humber comes close to advocating a system of strict liability for all criminal offenses. He does allow that a person may be acquitted if a crime was committed accidentally. (It is unclear, however, why he should make this exception. Is it easier to determine lack of intention than to determine lack of competence or responsibility?) Just as current strict liability laws are usually justified by showing that they protect the public or society, so Humber would justify incarceration for all serious offenses on the grounds that this policy would protect society. Convicted criminals are not to be punished because they have done something morally wrong that requires retribution. Nor are they to be punished as a deterrent to the commission of crimes by others. Indeed, one should not attempt to punish them at all, because, in Humber's eyes, the very idea of punishment is suspect. In its place he would put Richard Wasserstrom's notion of exclusion.[4] Criminals are to be excluded from society for a definite period of time so that society will be protected against them. The length of imprisonment is to be a function of the degree of seriousness of the crime—the greater the seriousness, the greater society's need to be protected and hence the longer the sentence. The system of sentences can be determined by the political process, so that the people of society, at least indirectly through their representatives, have a say in assessing the various threats to society. All sentences must be determinate; the only reason that could be given for indeterminate sentences is the claim that psychiatrists should ascertain when prisoners are ready to return to society, and Humber has challenged their ability to do this. Nonetheless, Humber does want psychiatric treatment to be available to prisoners, but only under very restricted circumstances. If a diverse group of psychiatrists can agree unanimously that a prisoner is mentally ill and incompetent, the prisoner may be hospitalized and treated, always with a view to rendering the prisoner competent again. Although we cannot be sure that treatment is in the prisoner's interest, under the conditions stipulated it is our best "educated guess" that involuntary commitment would benefit him or her. But to protect hospitalized prisoners against the possible ignorance and confusion of psychiatrists, they must be freed when their sentences are completed. It is in these latter respects that Humber's proposal differs from, and I think improves upon, Lady Wootton's recommendations. Although voicing doubts about our ability to determine responsibility, she has no such qualms about our ability to decide the right therapy for a convicted person and the right amount of it.[5] This position is arguably inconsistent, and Humber is wise to steer clear of it. The same doubts that led him to restrict psychiatric involvement in a criminal trial lead him to impose strict limits on the psychiatric treatment of prisoners. Within these protective limits, let the good doctors do what they can for a person.

I find Humber's argument to be interesting, challenging, and, in part, persuasive. Although a number of psychiatrists are certain to disagree with his assessment of the current state of psychiatric knowledge, a degree of skepticism surely is entailed by the disagreements within the psychiatric community. Indeed, a number of philosophers today, following the lead of such thinkers as Herbert Morris and Jeffrey Murphey, are even more critical than Humber of psychiatric involvement in the criminal process.[6] They see many psychiatrists operating with a medical model of the person that would supplant the concept of a person as a rational, responsible agent—a concept that is at the foundation of our legal system. At the same time, very few philosophers want to abandon psychiatry altogether. Hence a proposal such as Humber's that attempts a compromise between total acceptance and total skepticism should be considered carefully.

I do feel, however, that Humber's ideas contain a number of problems. I will indicate these and make some alternative suggestions of my own.

First of all, I am doubtful about the notion of exclusion (borrowed from Wasserstrom) that is to be put in the place of the idea of punishment. What we have here seems to be a distinction without a difference. Call it punishment or exclusion, the prisoner is denied liberty. This denial comes about as a consequence of his or her breaking a law, and it is administered by officials authorized by law to do so. But the notion of a deprivation imposed by officials for an infringement of law is precisely what many philosophers today take to be the formal definition of punishment.[7] Furthermore, it seems that the justification

Humber offers for exclusion is not altogether different from the justification of punishment given by the deterrence theory. To imprison a
lawbreaker is to prevent that person from committing the crime again
for the time limit of the sentence. What Humber gives us might be
thought of as a very restricted theory of deterrence. It is a theory, however, that escapes the usual criticisms brought against the deterrence
view, namely that imprisonment does not in fact deter others and that it
is wrong to punish a person in order to influence someone else. Clarity
and perhaps cogency would result if Humber called his proposal a limited deterrence theory of punishment.

   These may appear to be minor points, but I do not think they are.
If what we have here is a theory of punishment under another name,
then we need to see whether it too encounters some of the standard
problems of theories of punishment. The one that I wish to concentrate
on is the problem of the nonresponsible criminal. Humber himself admits that ''if mentally ill persons are not morally responsible for what
they do, they do not deserve punishment.''[8] It follows that if there are
nonresponsible criminals, and if Humber's theory is in reality a theory
of punishment, he would be condoning something he admits is morally
unjustifiable. His position, of course, is that we do not have the satisfactory ability to distinguish the responsible from the nonresponsible
lawbreakers. He also believes that in incarcerating lawbreakers we are
not really punishing so much as excluding them for the protection of
society. In response, we need to question whether Humber has properly assessed our ability (or lack of it) to determine responsibility.
More on that point later. We also need to recognize that, whatever you
call it, incarceration in our present penal system brings with it stigmatization, degradation, and great danger.[9] If nonresponsible persons were
to be incarcerated under Humber's scheme, they would be treated most
unfairly. He says that prisons should not be 'mean' places, but to trans-
form our present hellholes into congenial places would require budgetary and attitudinal changes in our society that are staggering and incon-
ceivable. Nothing could be more impracticable, a point that Humber
should appreciate.

   Another of my objections to Humber's concept of exclusion pertains to the determination of sentences. He claims to have offered us a
way in which determinate sentences may be arrived at in a nonarbitrary
way. Society is to judge the seriousness of the various kinds of crime,
and then it is to devise a system of sentences that assigns a definite
period of incarceration to each kind of crime. The greater the seriousness of the crime, the greater the period of incarceration. But what is
meant here by 'seriousness'? Apparently the degree of seriousness of a
crime is a function of the degree of the threat to society that the crime

involves. How, then, are we to determine the degree of the threat and what is needed to protect society from it? Humber suggests that the members of society, through their political representation, are to determine this. But such a democratic procedure likely would be no more than an appeal to the emotional reactions of the members of society to the various kinds of crime. What undoubtedly would eventuate is that the degree of social outrage would determine the seriousness of a crime, and hence the length of sentence for it. There may be nothing wrong with this notion *per se,* although I am disturbed by the thought of what degrees of outrage might greet homosexuality, or unmarried sexual conduct, or leftist political activity—some members of society would consider these worse than ordinary bank-robbery, especially if the robber were a white collar one. But aside from these concerns, let us ask what society's moral outrage has to do with the protection of society? Supposedly, exclusion is justified because it protects society, but it is doubtful that a sentence determined by the degree of moral outrage has much chance of being the best way to protect society.

I think that Humber is shortsighted when he says that "in order to assure the protection of society, the question of *why* someone refuses to accept society's rules is wholly unimportant." Understanding the state of mind, or the state of society, that led to the crime is crucially important if we wish to protect society to the fullest. Unless we understand the motivation that produced the crime or the circumstances that provoked it, we are in no position to change things either in the criminal or in society so that the criminal action will not be repeated. To be sure, we remove threats momentarily by incarcerating criminals, but unless we keep them in jail for the rest of their lives (an option Humber is unwilling to endorse), we run the risk that they will repeat their crimes when released and that they will pick up even nastier habits while in jail. One thing seems certain: society is not *necessarily* protected more (in the long run) simply by assigning a longer imprisonment. For all we know, a murderer may be changed more effectively by a moderate sentence than by an extreme one; a pickpocket may require more imprisonment than a murderer in order to change his or her ways. What we need here is empirical evidence concerning the consequences of imprisonment (and imprisonment in *our* prison system), and while such evidence is undoubtedly very hard to come by, Humber has not convinced me that it is totally unavailable and that we must turn the matter over to the intuitions and emotions of the average citizen.

Humber himself seems to hedge a bit concerning our ability to understand criminals. He does allow for psychiatric treatment of a criminal if a group of psychiatrists of different theoretical persuasions agrees on the diagnosis of the prisoner. To be sure, he says that he is working with an idealized model here. Does this mean that he expects the psychiatrists never to agree? If so, he is ruling out all involuntary forms of psychiatric treatment. But if he thinks they occasionally will agree, on which occasions they may impose treatment on an incompetent patient, then why not allow for the possibility that such a representative group of psychiatrists may be able to agree before or during a trial, in which case it would be reasonable to allow them to have input at that stage? And if we could get some agreement from them on the question of the effects of various lengths of sentences on prisoners, why not allow in these instances for a psychiatric determination of the appropriate sentence? I do not know whether psychiatrists *can* agree in these limited ways. I am suggesting only that if we may expect the agreement Humber talks about, we may equally expect such agreement at other points in the criminal process, and, moreover, that if we get this agreement, it would be only reasonable, using Humber's own argument, to allow psychiatrists to play a greater role in the determination of how we are best to protect society.

Humber might object that the determinate sentence "protects" the prisoner when psychiatric treatment is allowed in the post-sentencing stage, but that an accused or convicted person would have no such protection if psychiatric input were allowed earlier in the process. But why not introduce alternative protections at these earlier stages? If psychiatrists can agree that an offender was mentally incompetent at the time of the crime, then moral fairness demands that the offender be found not guilty. In that case, why not commit him or her to an institution for treatment and have a citizen's group review periodically the inmate's case, a review that would consider the reports from a number of psychiatrists and from the inmate? If the psychiatrists disagree over whether the inmate is cured, and/or if the inmate protests continued confinement, this board of citizens would have the right to terminate treatment or to set a definite time limit for continuing treatment. In this way the patient's rights would be protected, the stigmatization of a guilty sentence and the dangers of prison life would be avoided, and the opportunity to benefit from society's best educated guess about proper treatment would exist.

Frequently in his essay Humber appeals to the principle of justice that states that equals should be treated equally and similar cases dealt with similarly. He seems to think his proposal for determinate sentences for all criminals regardless of their mental states is consistent with and supported by this principle. In the scheme of things as he visualizes it, "a murder committed by a rational, free agent and a murder committed by a mentally ill person would be similar in all relevant respects; and society could not claim that it needed to treat some prisoners differently in order to protect itself.'' This is all well and good if we assume that Humber's proposed system of criminal law and procedure is already in operation. It would be a violation of justice to treat two persons differently if the law prescribed the same sentence for both. But if we are considering whether we should institute such an expanded system of strict liability, we need to remember the other part of the principle of justice as Aristotle gave it to us: justice requires that we treat unequals unequally.[10] It is wrong to respond in exactly the same way to two lawbreakers who are importantly different. The difference between a responsible and a nonresponsible person is an important, relevant difference. Finding both of them guilty and sentencing them to the same period of incarceration is to treat those who are dissimilar in similar ways, and that is a violation of justice. The suggestion I have just made for an alternative way of handling mentally ill lawbreakers, a suggestion that is consistent with Humber's skepticism toward psychiatry, attempts to avoid that injustice.

Finally, let us consider the question whether Humber has correctly assessed our ability to distinguish the sane and insane, the competent and incompetent. Much of what he says undoubtedly is correct, and there is no doubt in my mind but that psychiatry in some respects is in a muddled state today. But we need to remember that in considering the question of whether lawbreakers are responsible for their actions, we do not have to rely solely on psychiatric theory and judgment. As Humber notes, and most philosophers would agree, being mentally ill need not make one nonresponsible. It does so only if it generates an inability to act otherwise. This inability to act otherwise is the key concept from the standpoint of criminal law, and it is a concept we know how to employ fairly well in everyday life. We are familiar, that is to say, with forms of behavior and context that indicate an inability to do otherwise. Although often puzzled by borderline cases, we are not puzzled by the nonborderline cases that are much more frequent. Now psychiatry may reveal new causes of, and new forms of, nonresponsible behavior; but the core concept of being unable to do otherwise remains the same, and it is difficult to believe that we are completely stumped by all psychiatric cases, not knowing whether these individuals are able to do otherwise or not. What generates confusion in the case of psychiatry is that psychiatrists are attempting to do much more than determine responsibility; they are concerned with the etiology, classification, and treatment of a vast variety of diseases. And uncertainty with respect to these issues suggests uncertainty across the board, even with respect to responsibility. Indeed, many psychiatrists are confused about responsibility. Some of them wish to abandon the notion altogether, thinking it unscientific because it conflicts with psychological determinism or because, as Lady Wootton argued, it is something "under the skin" and beyond the view of the scientific observer. Neither of these claims is correct. There is available today a vast philosophical literature that canvasses the concept of responsibility and delineates the conditions under which we should, and in fact do, apply it.[11] If psychiatrists attended to this literature, they would find assistance in defining the general concepts of mental illness and incompetence. It is likely that they then would be able to bring greater clarity into the courtroom regarding the mentally ill offender. It is because we are convinced that many people break the law as a result of being unable to do otherwise, because we know that we ourselves and our acquaintances occasionally behave nonresponsibly, that we feel there is a serious moral problem with strict liability in the criminal law. To criticize psychiatry for not offering us the help it should is a salutary endeavor. To abandon a conception of law that is based upon the notion of individual responsibility is altogether something else, and in my eyes potentially dangerous. Let us not tie the notion of responsibility so closely to the notion of psychiatric knowledge that to question the latter involves jeopardizing the former.

Notes and References

[1]James M. Humber, "The Involuntary Commitment and Treatment of Mentally Ill Persons," *Social Science and Medicine* 16(F) (1981) pp. 143–150.

[2]Francis G. Jacobs, *Criminal Responsibility* (London: Weidenfeld and Nicolson, 1971); and Nigel Walker, *Crime and Insanity in England:* The Historical Perspective (Edinburgh Press, 1968).

[3]Lady Barbara Wootton, *Crime and the Criminal Law* (London: Stevens, 1963).

[4]Richard Wasserstrom, *Philosophy and Social Issues* (Notre Dame, IN: University of Notre Dame Press, 1980), pp. 136–137.

[5]Lady Barbara Wootton, *op. cit.*, 177.

[6]Herbert Morris, "Persons and Punishment," *The Monist* 54(4) (1968), pp. 475–501. Jeffrey Murphy, "Criminal Punishment and Psychiatric Fallacies," and "Preventive Detention and Psychiatry, *Punishment and Rehabilitation* (Belmont, CA: Wadsworth Publishing Co., 1973).

[7]A. Flew, "The Justification of Punishment," *Philosophy* 29 (1954), 291–307; S. I. Benn, "An Approach to the Problems of Punishment," *Philosophy, 33 (1958), pp. 325–341; and H. L. A. Hart, "Prolegomena to the*

Principles of Punishment," *Proceedings of the Aristotelian Society,* 60:1–26 (1959–60).

[8]Humber, *op. cit.,* 145.

[9]Hugo Adam Bedau, "A World Without Punishment?" in Milton Goldinger (ed.), *Punishment and Human Rights* (Cambridge, MA: Schenkman Publishing Co.: 1974, pp. 141–162).

[10]Aristole, *Nicomachean Ethics,* Book V.

[11]See Bibliographies in R. Binkley, R. Bronaugh, and A. Marras (eds.), *Agent, Action and Reason* (Oxford: Basil Blackwell, 1971); and Myles Brand (ed.), *The Nature of Human Action* (Glenview, IL: Scott Foresman and Co., 1970).

Section V

Patenting New Life Forms

Introduction

There are a variety of issues emerging from the recent trend on the part of researchers to patent and commercially exploit newly developed microbiological techniques. In the first part of his essay, "Patenting New Forms of Life: Are There Any Ethical Issues?", Professor L. B. Cebik introduces the reader to some of the most important of these issues. In the second portion of his article, Cebik selects the following four issues for ethical analysis.

First, Cebik considers the call for a new Asilomar conference to set guidelines for biologists engaged in the commercialization and patenting of new discoveries. Cebik concludes that although a conference of this sort would not succeed in solving any problems, it might serve to more precisely identify and formulate problems that researchers will face as they attempt to commercially exploit their discoveries.

Second, Cebik considers the claim that patenting and marketing by university research personnel will destroy the public image of science and the universities. Cebik claims that the present image the public has of science and the universities is a myth; and he contends that this image ought to be changed so as to better reflect reality.

Third, Cebik examines the opposition between the demands of science and the demands of profit: science demands openness and sharing of information, profit requires secrecy—at least until protection is assured via patents. At present, Cebik contends, we have no way of adjudicating this conflict, for we have no standard by which we can give one value priority over the other.

Finally, Cebik considers the claim that the Supreme Court decision to allow patenting of new life forms will alter our understanding of life and lead, inevitably, to a devaluation of human life. Cebik rejects this view, characterizing it as an irrational fear.

In "Ethical Issues Raised By The Patenting of New Forms of Life," Professor James Muyskens raises a number of questions concerning the propriety of university research personnel patenting and commercially exploiting new life forms.

First, Muyskens questions whether we ought to permit patenting and commercial development of new life forms. Muyskens points out that those who favor patenting and commercial exploitation seem to assume: (1) that new life forms are human inventions, (2) that university research personnel are the inventors of the new life forms they create, and (3) that university research personnel (inventors) have a moral right to control their creations. Muyskens questions these assumptions, and argues that patenting and commercialization ought to be permitted only when it is clear that society-at-large (especially the least advantaged members of society) will benefit from such action.

Second, Muyskens points out that a number of problems evolve when university research personnel become employed in private industry while at the same time retaining their university positions. Muyskens argues that in these cases researchers lose objectivity, research quality suffers, and conflicts of interest arise. After considering a number of alternative suggestions, Muyskens contends that some compromise on the principles of secrecy (as required by industry) and the free flow of information (as required by university based research) is needed. As an example of an acceptable compromise, Muyskens cites an agreement by Yale University researchers to keep their findings secret for 45 days when requested to do so by private employers. After 45 days, all restraints on publication are lifted and researchers may publish their data.

Finally, Muyskens acknowledges that some may claim that his desire to place constraints upon university–industry connections is premised on a somewhat naive view of today's universities. Muyskens does not deny that contemporary universities have formed numerous alliances with government and industry. However, he argues that these alliances often were made on the basis of expediency and were not intended as precedents. Furthermore, with the development of new life forms and the rush to commercially exploit such items, Muyskens feels we are now in danger of totally changing the way in which university research traditionally has been funded and conducted. And for Muyskens, change of this sort, though possibly beneficial in the short-run, would be detrimental in the long-run to both university research and industry.

Patenting New Forms of Life

Are There Any Ethical Issues?

L. B. Cebik

In the last year or two, some of the biological sciences stepped from a protected adolescence into the chaos and confusion of the adult work world. General Electric succeeded finally in patenting not only the Chakrabarty process for generating an oil eating microorganism, but the product—*Pseudomonas aeruginosa*—as well.[1] The Cohen-Boyer patent for their process of cleaving and recombining plasmids soon followed, with high expectations of an early patent on products.[2] These cases focused both academic and investor attention upon the commercial prospects of genetic engineering, now greater than ever, and consequently upon a number of new companies: Genentech, Cetus, and Biogen, to name three in the news. Harvard unsuccessfully proposed to its faculty that the university move beyond its more usual royalty sharing role in new developments to a minority stock share in a company specifically established with outside capital to exploit the patentable biological discoveries of one of its faculty.[3] Almost simultaneously, the University of California filed suit to protect its interests in a line of research said to have begun at UCLA and now placed under contract to Genentech by the commercial firm Hoffman LaRoche.[4] Surrounding these landmark events are changes in law and regulation to make it easier for universities to claim and market patentable ideas produced under federal grants and contracts and to lessen the stringent operating and reporting requirements for laboratories involved in recombinant DNA research.[5] As more universities, biologists, and businessmen flirt with the opportunities, the scenario will rapidly expand to the point of needing computer techniques to track the issues and the cast of characters.

The social issues that are emerging from the patenting and exploitation of recent microbiological techniques and products rival the DNA molecule itself for elusiveness, intriguingness, interrelatedness, and sheer complexity. Little wonder that some commentators have called for a new Asilomar to set guidelines for the involvement of biologists in the commercialization of discoveries pursuant to research in recombinant DNA techniques and whatever may follow as the next steps in our efforts to penetrate the molecular chemistry that undergirds life.[6] The call, however much misplaced, reflects a growing frustration and desperation. Is there an orderly means to grapple with the intricate questions facing universities, faculty researchers, investors, and a generally cautious American public?

A truly comprehensive analysis of the issues and their connections would begin at book length and grow from there. There are issues enough for law, for politics and government, for business and industry, for academic administration and faculty, and for ethics. One might be tempted to see the entire collection of issues as ethical. Nonetheless, ethical reasoning is likely to determine future events for only a few of the questions. Perhaps it may be of some use to sort and categorize—as best possible in a moving situation—the issues at hand, and then to select for further treatment those issues to which ethical analysis will most likely contribute in the coming years of discussion and contest. To achieve even such a limited goal, it is necessary to forego the usual introductory essay on the techniques of recombinant DNA. Our interest will focus on human concerns engendered by the patenting and commercialization of recombinant DNA techniques. The end result of this inquiry may add a word to the ongoing process, but it will be far from the last one. Indeed, in this new arena, we have yet to make a very good beginning.

Sorting Out Some Issues

Laboratory biology began its painful growth into social adolescence with the spread of research into recombinant DNA. Previously, only those portions of biology most immediately wedded to medicine, human or veterinary, felt much of the impact of the complex entanglements of commercial and social ventures, and medicine largely enveloped the aspects of biology it most needed. Academic biology carried on its sequestered life of producing knowledge for its own sake. Pubescent self-awareness emerged in the shadow of dangers, both real and perceived, of delving into the possibilities made theoretically feasible by DNA theory. *E. coli* was to become a more familiar name than those of the researchers who practiced the techniques of gene splicing. For the first time in this century, biologists grew extremely self-conscious as they reflected upon the potential harm of their work.

Asilomar, that 1975 conference at which scientists grappled with their responsibilities to protect the public from hazards, whether intentional or accidental, growing out of their research, largely restored biological self-confidence. Academic biologists returned to their laboratories confident that their distant colleagues would adhere to the same rigorous standards they adopted. Few of them dreamed that they had taken a first major step toward the issues to erupt suddenly in 1979 and the present.

The Patent

General Electric's successful appeal to the Supreme Court in *Diamond, Commissioner of Patents and Trademarks, v. Chakrabarty* to permit patenting of laboratory-created microorganisms occurred in 1980. However, a year earlier, with the decision of the US Court of Customs and Patent Appeals to permit patenting of new organisms from the laboratory, serious questions began to emerge. The most immediate issues to arise from the decision, in which the Supreme court appears to echo largely the lower court ruling, surrounded the flow of information among basic researchers. Idealistically, some researchers looked at their work as open and available to all, and expected as much in return. Others appeared to believe that some current work is being sequestered like industrial trade secrets for fear that leaks might reduce the chances either for profit or for honor. Patents, so it has been argued, protect while making information available for researchers and others without commercial motivations.[7] Ferreting the statistical truth of these claims might be an idle exercise. Biological science, like chemistry, physics, engineering, and a number of other fields in modern academia, had already stepped across a line that made the ideal of free knowledge for all and its own sake merely nostalgia. Commercial linkage with academic biologists, as with so many others within universities, had made life evermore a matter of acceptable compromise between entrepreneurship and scholarship, all depending upon the individual aspirations of the researcher and the rules of his or her university.

More profound, although perhaps no more or less serious, are questions about the nature of life and its place both in the law and business. The Supreme Court decision takes patent law to cover "human-made inventions" in contrast to unpatentable "products of nature."

Inventions might include living materials as well as inanimate objects. The Court also noted that Congress was free to alter the law to be as restrictive or broad (within limits) as it might choose, thus making the present ruling a legally narrow one.[8] However, the lower court decision tends to accept the view that there is no effective way to distinguish between the living and nonliving at the level of microorganisms relevant to the case. The distinction between the living and nonliving has a long tradition in western thought. (Some philosophers, such as Descartes, have, of course, challenged the idea in one or another way.) The Court's ruling appears in some eyes to sanction dismissal of the distinction. Life *per se* ceases to have a sacred aura. The only remaining bastion to protect the specialness of human life seemingly lies in the idea of being human, and this, too, is not without its attack from some quarters.

Commercialization

Regardless of its ultimate impact upon our conceptions of life and human life, the prospects for profit from the biological techniques developed in university laboratories have given rise to a number of new companies in the private sector, all planning to exploit new genetic technology. Corporate names such as Cetus, Genentech, Genex, Hybritech, Biogen, and Bethesda Research may never become household words, but these small firms are taking their place beside innumerable engineering, chemical, and other counterparts devoted to profit from commercial applications of scientific findings and techniques. Academics or recent academics head most of these firms on the technical front, and increasingly on the business front as well. As these companies emerge and develop, it is common for the academic entrepreneur to split his or her effort between a full-time university position and whatever effort the fledgling firm requires, generally with the blessing of the institution. In the eyes of administrators of schools experienced in research that is accompanied by skill or product spinoffs, such arrangements are mere extensions of consultation policies ensconced within institutional manuals.

For whatever original motives, universities with significant research and professional programs almost universally provide faculty with permission to consult at least one day per work week with entities outside the university. Whether this policy is necessary still to attract and retain first quality people is for some an idle question: faculty expect as a matter of course what in other fields and employment circumstances might be called moonlighting. The fact that the practice applies to those faculty with salable skills and not usually to scholars in the traditional senses (e.g., the humanists) has sometimes yielded dispute within universities over whether the policy is in fact a privilege not open to all and hence unfair. However, quietude on the matter generally pertains. Indeed, faculty in nonconsulting fields often are unaware of consulting policies.

The development of new biotechnical corporations has raised some of the issues surrounding both the consulting and entrepreneurial activities of faculty from a fresh perspective. The heavy investment in Genentech as its stock was announced suggested that faculty in this new area might reap heavy profits through commercialization efforts. Whether faculty members could effectively sustain proper loyalty to a university under conditions of conducting research in two separate, but interrelated institutions remains an unanswered question troubling some of their colleagues. Potential competition with the university has also been envisioned. It is not yet clear to what extent these are in fact real problems for any particular university and to what degree they are point-winning claims in faculty debates.[9]

Conflict of interest laws and their application within the various states provide little guidance either to faculty or to universities, whether public or private. If practice seems to wallow in vagueness, the appearance reflects a reality wider than university boundaries. Numerous professions accept moonlighting as a matter of course; in police and fire protection work, moonlighting is often considered normal to supplement incomes and to utilize lengthy off time periods created by scheduling that is foreign to the eight-to-five world. In the absence of applicable standards from any source, the issue will likely remain one for local negotiation within the limits created by state law, federal regulations (including accounting rules), and institutional tradition.

The University Company

Harvard's brief foray into the possibility of joining with outside capital as a minority stockholder in a biotechnical company introduced a new element into the entrepreneurial formula. Harvard would license the company for exclusive use of a faculty member's patentable discoveries and collect profits directly. Unlike more common royalty arrangements, Harvard would maintain some control over that to which its name was attached, but as well sustain some capital risk in the event the company encountered difficulties. As Barbara Culliton has noted, the primary motivations for any university directly entering the marketplace, rather than relying heavily upon federal grants and contracts, are three: to acquire new sources of income in the face of federal reductions, to obtain some freedom from federal regulatory requirements (especially accounting and audit mandates, but numerous others as well); and to obtain freedom from federal control of research directions (even the seemingly freest grant programs fund in clustered directions since money for research is far from unlimited).[10] Of the three, the first has become almost a matter of desperation in the face of rising costs and leveling, if not decreasing, funding levels in most of the sciences.

Harvard retreated from its proposal in the face of negative response from its faculty of Arts and Sciences. As *Nature* reported, "The university administration seems from the beginning to have been keen on the prospect of an alternative source of income. If it had seemed less greedy, the Faculty of Arts and Sciences might have been less quick to take fright. But in the end, the proposal foundered on the faculty's misgivings. For a time, Cambridge (Massachusetts) seems to have been thick with anxiety about secrecy, academic freedom, and the like, but perhaps the clinching argument against the comercial venture was that the university could not, in the arrangements proposed, avoid discriminating in favor of those on whose research its financial prospects seemed to depend. In a college of supposed equals, some would be more equal than others."[11] Harvard Medical School faculty member Bernard Davis found the Faculty of Arts and Sciences perhaps lax in their sense of the "long history of negotiations, especially in medical schools, over academic activities and positions linked to private gain" as they arrived at their veto of President Bok's proposal.[12]

Perhaps more than the history of individual private gain, the faculty of Harvard, and most other people as well, are ignorant of the development mechanisms that have allowed universities to attract funds and profits. Many writers seem to be under the impression that patent royalties represent the only applicable model for nonstandard university income. It is probably more nearly correct to suggest that within the limits of image and law, most universities are already engaged in using profits (or excess of income over costs) from one operation to support other operations within the institution. Much of this is simple budgetary readjustment and is equally applicable to both private and state supported universities. Hardly a university in the country does not have a development office whose mission it is to generate from alumni and other donors significant income, the less earmarked for specific purposes, the better. Scholarship funds, library funds, and a host of others benefit from efforts that sustain the nonprofit status of universities.

University research efforts have developed a number of other means to maximize their income in support of faculty work. Mandatory patent and copyright policies are becoming more the norm. Universities are beginning to demand first rights for shared royalties from the work of their faculty performed on university time and facilities. The pattern of such policies—and their terms—is far from universal, and some major research institutions still have no policy at all in this area.

The means by which universities administer patent and other research policies also vary widely. In public institutions, state law may limit the freedom available for engaging in some income-producing activities, whether related to commercial enterprises or to grant and contract work. Private, nonprofit research corporations have sprung up around the country (and in some places died as well). The scope and activities of such corporations vary even more widely than patent policies. Some are content to accumulate patent royalties and other safely produced income in the form of certificates of deposit. Such conservative financial attitudes contrast to other schools whose research corporations generate venture capital to invest in university related research. The level of activity and the placement of efforts largely stem from the fiscal aggressiveness of a university's central administration, as well as whatever boards, legislature, or other policy-making bodies stand above the institutions. Few academic discussions of the Harvard case, or similar ideas broached in Michigan and other places, have given much place to these varied institutional mechanisms to enhance income and research efforts. In fiscal matters, faculty and even many administrators exhibit a naiveté with respect to practices common throughout the country, leaning rather more heavily upon an image and an ideal that are no longer fully operative (if in fact they were ever fully operative among institutions, whatever the high standards sustained by individual scholars and scientists).

Laws and Regulations

The Harvard faculty refusal to countenance willingly a university involvement in a private profit-making corporation comes in the face of relaxing regulations regarding the conduct of research involving recombinant DNA and laws encouraging universities and small businesses to develop and market ideas generated under federal grants and contracts. The same issue (and page) of *Nature* that reported the Boyer-Cohen patent also reported the legislative passage of the bill to permit uniformly throughout government agencies retention by universities of patent rights of discoveries made under federal grants, so long as proceeds return to the support of teaching and reseach. Agencies such as the National Science Foundation and the National Institutes of Health (under Health and Human Services) had already been striking Institutional Patent Agreements. The entire effort was designed in part to promote technology transfer, thus shortening the route of scientific discovery into the marketplace and to the public. Agencies retained, interestingly, the right to direct such licensing operations should universities fail to pursue their option vigorously.[13]

Since the institution of NIH Guidelines for Research Involving Recombinant DNA Molecules in 1976, accumulated evidence has suggested that some, if not most, of the fears among scientists regarding dangers to humans and other organisms of both the beginning elements and the final products of laboratory work with DNA have not been substantiated, despite intensive work and specific experiments to tests dangers.[14] Consequently, for at least two years, NIH has taken steps to reduce the degree of control over such experiments and to reduce as applicable the required precautions of investigators. Early in 1981, two members of the Recombinant DNA Advisory Committee suggested conversion of current guidelines, which have the force of regulation, into a volunteer code. The motion comes only six years after the Asilomar conference. The entire Advisory Committee found the move too drastic, promising to study it for six months and to continue reductions of controls on a warranted piecemeal basis.[15]

The directions taken by the federal government generally encourage the development of both research and commercialization of recombinant DNA work. In some quarters, marketable applications of such work are perhaps already overdue. Several readily conceived applications are envisioned in the Boyer-Cohen patent application: "The method provides a convenient and efficient way to introduce genetic capability into microorganisms for the production of nucleic acids and proteins, such as medically or commercially useful enzymes, which may have direct usefulness, or may find expression in the production of drugs, such as hormones, antibiotics, or the like, fixation of nitrogen, fermentation, utilization of specific feedstocks, or the like."[16] Genentech has been at work developing methods to produce somatostatin, a brain hormone, and insulin. Its contract with Hoffman LaRoche for work on the production of interferon is being contested by the University of California. At the September, 1981 meeting of the Recombinant DNA Advisory Committee, The Committee announced approval of four different requests from Genentech, Cetus, Schering, and Molecular Genetics for large-scale culture work related to DNA.[17] With a growing number of companies entering the field, these publically announced efforts represent but a small part of the overall effort, which—in the absence of radical changes in laws or regulations—is likely to grow steadily in the future.

Part of the confusion surrounding efforts to patent and exploit commercially the biotechnology of recombinant DNA stems from a failure to recognize that federal policies, whether blindly or cognizantly, have produced a growing sanction in their favor. To the extent that governmental action reflects the will of the national community, then commercial development of recombinant DNA techniques and products already has its mark of general approval. (Of course, specific techniques, products, and practices must still pass muster on a variety of legal, fiscal, ethical, and other grounds.) The net effect of claims that would halt virtually all such work is then either to suggest that governmental actions in fact do not represent the national community or to assert that there may be overriding reasons for superceding the national community. Daniel Callahan argued in 1977 that the public had been insufficiently involved in the debate to that date.[18] At the same time, Tabitha Powledge described the employment at a public Academy Forum of such "very technical language that effectively, if unconsciously, shut out most of the public."[19] Nonetheless, it remains the fact that as of 1981, the public voice has not shown itself significantly divergent from the tide of events. Consequently, whether there are reasons for superceding governmental action in behalf of the national community and what such reasons might be is at this state of discussion quite unclear: one more reason in itself for noting that we are still at the beginning of the debate.

Among the specific issues that deserve attention is whether the NIH guidelines are in fact adequate, and whether NIH is the best agency to administer them. Key Dismukes, in his 1977 review of the guidelines, noted that "the NIH approach fails to meet two critical requirements for regulation of recombinant DNA: enforcement of industrial compliance and monitoring health."[20] With regulations four years later moving toward a more voluntary basis with respect to institutions where enforcement had been possible, the prospect for enforceable rather than voluntary adherence to the guidelines by large industries is less likely in the short run. Dismukes proposed an EPA style approach to the problem, with oversight by the Public Health Service. Whether or not this suggestion has final merit, it does mark the continuing negotiation and dialog on the adequacy of protection measures and the fundamentally political, technical, and public nature of the specific questions.

Some Ethical Issues

The brief catalog of issues has oversimplified the interconnections among them for the sake of clarifying certain facets of the discussion. What emerges with some distinctness is a pattern of public negotiation among varied parties as the direction of commercializing the fruits of DNA research evolves. Not all parties to the discussion have yet made themselves known. Despite reasonably wide press coverage, not only in scientific media, but in more general news media as well, much of the public has yet to formulate views on the impact of these develop- ments. Before the discussion has reached durable conclusions, reli- gious, environmental, and similar groups very likely will take posi- tions regarding either the short- or long-range implications of both the general expansion of recombinant DNA work and specific facets of im- plementation, beneficial or detrimental. It is far too early in this ex- pression of the American ethos to forecast outcomes.

Nonetheless, there are a few issues that deserve more extensive comment. Among the questions around which we might focus discussion are these: the proposal for a "new" Asilomar conference, the public image of science and the universities, secrecy and profit, and the status of the concept of life.

The Call for a New Asilomar Conference

The now famous Asilomar conference, held in February, 1975, culmi- nated two years of discussions among biologists over the safety of con- ducting recombinant DNA research. Historically, the Gordon Confer- ence of 1973, from which the Asilomar conference emerged to answer the issues raised, might qualify as the more significant; nonetheless, the California event that recommended the lifting of the moratorium on recombinant DNA research, as well as NIH guidelines for its conduct, has gained the greater prominence. As a result of its success, as meas- ured retrospectively, the suggestion of a second Asilomar has been voiced, perhaps first by Stanford president, Donald Kennedy.[21] There seems to be a parallel between the pressures of commercializing bio- logical research and former concerns about the safety of recombinant DNA research in the variety of problems scientists face that they have or had not previously faced. Thus, the model of Asilomar seems *prima facie* applicable.

Culliton, while admitting that post-Asilomar I debate was "some- times acrimonious, sometimes downright foolish, and often tedious," still concludes that "an 'Asilomar II' might not be such a bad idea. If a conference helped to clarify the issues and set out the choices, it could not help but to contribute to resolutions. However, it is not obvious that a single codified set of rules is needed. Instead, if certain general principles could be agreed upon, the door might be open to a variety of arrangements that could be tailored to individual institutional and industrial needs.''[22] Although this argument is certainly appealing, and indeed an Asilomar II should be held if for no other reason than to garner a full and fair exchange of views, the relatively instant and full success of Asilomar I is not in the cards, and certainly not without either a widening of the range of participants and/or a great deal of nonbiological homework by the participants. The conclusions of Asilomar I came to fruition in the offices of NIH with the advent of the Guidelines for Research Involving Recombinant DNA Molecules and allied structural elements, such as the Institutional Biosafety Committees. These regulatory actions, hard on the heels of Asilomar (as governmental actions go), were possible in part because of the continuity between biologists in the field and the biologically well-trained personnel at NIH. Quick, enforceable action, applied with relative uniformity across the field, provided the Asilomar conference with its halo of success.

An Asilomar II to explore the problems of patenting and commercializing of DNA research results by universities and/or university biologists stands less chance for equal success for many reasons. First, in the arena of commercial involvement by university scientists, there are few standards. The divergence of involvement by engineering and scientific colleagues within universities provide biologists with few useful models. Second, in the arena of commercialization, there is no agency comparable to NIH to supply quick, enforceable action subsequent to an Asilomar II. Commercial ventures, whether by individual faculty or by universities will be governed by applicable law and regulation, responsibility for which is split among a large number of agencies. Responsibility especially passes from the understanding confines of NIH to more generally engaged offices for whom biologists are but one of a myriad of types of people doing business.

Culliton's suggestion that the development of general guidelines permitting many options for individuals and institutions as a third reason not to overanticipate an Asilomar II may yield vacuous results. At worst, such loose guidelines would be self-contradictory, guiding nothing by allowing too much. At best, they would still lack enforceability. Even censure of individuals or institutions for violation will carry little weight compared to the success or failure of the particular venture involved. The utility of censure may be limited, since many

individuals will be guided, if not led, into commercializing mechanisms countenanced by a particular institution rather than disciplinary
codes. As noted earlier, the range of fiscal activities conducted by an
institution directly or through attached nonprofit research corporations
or similar entities varies widely for reasons of both law and characteristic business practices of boards and administrators. If an Asilomar II
does not include a fair sampling of research administrators and central
officers of universities, it will lack sufficient expertise even to identify
usefully the problems to which it wishes to contribute. The alternative
is for participating biologists to take a crash course in fiscal, research,
and central university administration as it may relate to problems of
commercializing the results of recombinant DNA research.

Discussions of commercialization of biological research are inherently more broadly based than discussions of the safety of certain orders of the research itself. The patenting of research results and the
growth of new companies that intimately involve academics precludes
biology from achieving a solution in isolation. Isolated action may in
fact create new competing pressures rather than ease or eliminate
them. Although such situations may be inevitable in the modern world,
intentionally and avoidably contributing to them through the creation
of isolated standards may not itself be in the best interests of rational
practice for any discipline involved in research. Notwithstanding these
difficulties, a broadly based Asilomar established for the more conservative goal of identifying and formulating the problems of commercialization from the perspective of the scientist stands some chance of
success and likely would not contribute to the complicating of the
problems.

The Public Image of Science and the Universities

Closely related to concerns that have prompted the idea of an Asilomar
II is the idea of maintaining the proper public image of science, understood to mean the image of science as it is conducted within modern
universities. Fear that the free flow of knowledge among researchers
and from researchers to the public, as a sacred trust of universities,
might be jeopardized by excessive involvement in commercial enterprises has informed many a biologist's reluctance to see either
Harvard-style or individual entrance into the patent and marketing
world. In her *New England Journal of Medicine* piece, Culliton cites
Merton's 1942 four norms of science as a social structure: "universalism, communism, disinterestedness, and organized skepticism."[23] To
a large degree, the general public believes these features to be accu-

rately descriptive of not just biology or science, but as well of the entire university.

Perhaps one can only wonder whether it is ethical to permit the public to continue to believe such a myth. It has persisted despite open press coverage of grant and contract awards and activities of university faculty and departments. It has persisted despite the furor raised by many young dissidents over military and other security or proprietary research projects carried on at universities. Although many a faculty has indeed voted to prohibit such work, it continues in many universities. The myth persists despite longstanding involvements by faculty with commercial, profit-making ventures as spinoffs of research or professional skills. Most fields within the modern university are as involved in grant, contract, or commercial ventures as individuals within the departments desire.

This portrait is intentionally overdrawn to provide a first-order counter-agent to the effects of the persistent myth of free knowledge, freely communicated and freely given as the *sole* involvement of universities and their faculty. Universities are intimately bound to the total fabric of research and development within this country, although from some perspectives, they are not so bound that further ties are not possible. Much current work is externally directed through grants and contracts that universities and their faculty seek for a variety of motives, including money. The Harvard proposal was not novel except in its setting and publicity; the idea has been discussed in many settings throughout the nation, and in some cases, only legal prohibitions (or lack of marketable research results) have prevented action upon it. Too, European models have been extensively cited in the literature.[24]

For a number of years, university presidents have routinely paraded some of their involvements in various addresses aimed to show how universities contribute to the community, state, and nation through research and service, as well as teaching. Still, the purity of the university as a nonprofit educational organization beholding to the generosity of the public through either taxes or private giving persists. Too, science persists in its mythical image of purity, despite revelations of competition, secrecy, and commercial involvement. Perhaps it is true that myths persist because their believers will them to persist. If so, then these remarks will be wholly disbelieved or accounted as making a great furor over minor deviations from the norm. Certainly nothing in these facts affects the dominant attitude of universities to contribute as freely and self-determinantly as possible. Too, nothing can shake the dominant attitude of university researchers to conduct their work according to standards that exceed Merton's norms. But just how dominant these attitudes are and are likely to be in coming years, as pressures for further commercial involvements mount for motives of institutional need and personal profit, requires a hard empirical look. Should the image of either or both science and universities change in the process, perhaps it might be better to know this fact in advance so as to control the changes beneficially.

A change in the public image of science or universities need not entail a change for the worse. In the events surrounding the patenting and commercialization of recombinant DNA research results, no one has consciously proposed that individuals or universities engage in illegal or unethical activities. Fears and concerns attend what might result, even inadvertently and despite the best of motives. The Harvard faculty apparently did not want its administration even tempted toward discrimination on the basis of income, even though safeguards might be put into place.[25] Perhaps the sudden presentation of such an idea without detailed thought as to what images might become, or be made to become, has magnified the fear of negative results. Nonetheless, understanding what the university and science are in practice, as opposed to idyllic myth, might better permit a greater degree of control over the future. And it just might be fairer to the general public for which science and universities ostensibly exist.

The image of the science of biology and of the university as an institution cannot be separated from the realities of operating either enterprise. Both basic science and universities prize their independence of pursuit after knowledge. In recent times, both have felt assaulted by numerous pressures. For universities, burgeoning regulations make more difficult than ever the precarious balance between free inquiry and compliance. Many of these regulations also directly impinge upon the conduct of science, and biology—with little experience—now feels impending pressure from commercialization. In the face of these pressures, a value to be preserved in this world of ever more complex entanglements—however necessary and for the overall good of the nation—is the idea of free inquiry and the pursuit of truths wherever the trail may lead. David Smith, in his review of the conditions warranting the restriction of this right and goal, grounded only three: knowledge whose potential for abuse is highly probable, knowledge obtained and disseminated in immoral ways, and knowledge destructive of persons.[26] Perhaps much of the frustration facing research biologists in universities stems from the fact that encroaching pressures from many sources—regulation and commercialization being only two—may threaten the exercise of the right of free inquiry more than direct attack upon recombinant DNA research on grounds even remotely resembling Smith's.

Secrecy and Profit

The changing image of the sciences and universities, while a matter requiring attention, does not itself address the propriety of individual scientists withholding information they have developed in order to receive profit from the commercialization of those findings. Unfortunately, the manner in which the issue has been so far addressed obscures rather than clarifies the ethical issues involved. As presented in the press and in most other contexts, the question has been almost statistical in tenor: will the patentability of new microorganisms yield greater or lesser secrecy among scientists on the verge of discoveries?[27] Any answer to such a question as this must report on net gains or losses in the free flow of information for an entire field of endeavor. If microbiology winds up with a restricted flow of information, then it would appear that a socially desirable goal has been thwarted and hence social action—in the form of codes of ethics or the like—may be in order.

As formulated, the question of secrecy of scientific information has more a political tone than an ethical ring. However, most commentators have found it difficult to grapple with the question of secrecy on moral grounds. One can ask if keeping one's scientific findings secret is compatible with a scientific ethos that demands communal effort and free communication among colleagues, but no one does because the answer is so obvious. The contradiction resides within the question itself. Were free communication of biological techniques and findings the only question at stake, then the answer would be simple and the problem solved.

However, the problem involves more than the scientific ethos. For example, Dismukes, in commenting upon the Diamond v. Chakrabarty decision, expresses his concern for both "the secrecy with which genetic engineering is applied in industry" and "whether it is appropriate for scientists to profit from commercial applications of their research paid for with public funds."[28] What is at stake is a conflict of values, both of which are highly prized within our culture. Publication of research results for free use by one's colleagues is a longstanding component of scientific ethics. Indeed, no one has dared to challenge it, arguing instead that patents might make available information that some scientists are now holding as "trade" secrets in the hopes of profit. Profit is the other value in the conflict. Congress has settled many of the questions about the relationship of the funding of the research that brought about potentially marketable ideas and the actual profit made by researchers. What remains to be given are some reminders about profit as a value.

Individual ambition and acquisitiveness hold high position among American values. Within the university context, they often seem out of place, or at least not openly mentioned with tones of approval; nonetheless, free market profit motivation still provides one of the strongest mechanisms for realizing American self-esteem. In acquiring the scientific ethos through indoctrination and practice within biology—or any other science—an individual does not always cease to set aside the American dream. The increasing interaction between universities and other portions of American society has effectively destroyed the idea of the campus as an isolated bastion of nonprofit pursuit of the truth. The very activities that give the university its reputation for service also introduce to the classroom and laboratory an involvement in the values of the community that the institution serves. Thus, profit motivation by campus research scientists naturally becomes a more deeply felt force, although one that is held to different degrees by various individuals, depending upon their adherence to tradition and their resistance to changing influences. That some should be prepared not only to think about profit from their discoveries, but as well to act upon that thought in their own interest (and not just in the interest of the institution) is expectable, and more surprising by the slowness in the growth of numbers of those having such feelings than by the size of the phenomenon.

Profit, ambition, and acquisitiveness—within the bounds of socially acceptable practice—are positive values in this culture. That the genetic sciences as a whole do not yet see them so stems in part from the academic isolation of the past and in part from the fact that these values require actions quite at odds with the conventional requirements of values held within the scientific community. Whereas science demands openness, profit requires secrecy until protection is assured (e.g., via patents). But because two positive values are in operation here, no blanket praise or condemnation is possible regarding actions that appear to be in opposition to either of the values. At present there exists no standard by which to give one priority over the other. For example, to insist that a genetic engineer assign all profits from an invention to his or her institution would be no more and no less reasonable than to insist the same of an electronics engineer. In the dialectic of value conflict, there may one day be a reliable guide, one that affects all academics who face the conflict, and not just biologists. Until then, the decision as to which value takes priority will be largely a measure of individual judgment, desire, and commitment. It will also be a measure of institutional policy, law, regulation, and a myriad of other influences upon public values. For now, whenever another biologist enters the profit world, we can expect to see hearty doses of envy by those of like mind and resentment by those for whom the communal effort of science is paramount.

The Conception of Life

Perhaps the most fundamental issues aroused by the court decision to permit patenting of laboratory produced microorganisms concern our conception of life and the living. In fact, these are not new issues. Research involving recombinant DNA long before 1979 had created fears of abuse, especially with respect to eventually altering what it is to be human. Science fiction added hordes of off-human biological creations to whet the fear more finely, although many of these had little to do with DNA. Indeed, science fiction and allied genres of literature had long before created post-human beings through biological accident, radiation disaster, and simple, careful cross-breeding. It took first the prospect and then the reality of creating a new microorganism to raise speculative fears of violating an unspoken taboo to the level of public and private protest. Although the present state of the art makes significant genetic alteration of mammals a distant and uncertain prospect, the fear persists, abetted by some peculiarities of the court decision.

The challenge to the GE–Chakrabarty application by the commissioner was that it attempted to patent living organisms, something the law seemed not to intend. The Court's ruling drew a different line: between "products of nature, whether living or not, and human-made inventions."[29] To the extent that the Chakrabarty microorganisms were human-made inventions, they were patentable. At just this point disputes have arisen, arguing that the Court has attempted to draw a line between the natural and the human-made, thereby creating legal absolutes where it has no province. "How will the Court avoid interpreting what nature can and cannot do?" asked Lee Ehrman and Joe Grossfield, biologists writing in the Hastings Center Report.[30] Dismukes protests in the same issue that "Life is Patently Not Human-Made," and that "Chakrabarty did not create a new form of life; he merely intervened. . . ."[31] The Supreme Court may reason as it chooses, but these initial protests can be easily dismissed with a disclaimer that no absolute categorization is implied by the Court's decision. What is necessary to determine whether an entity, living or inanimate, qualifies as an invention is whether there has been sufficient manipulation, directly or indirectly (e.g., via machinery or designed chemical processes), by human beings to qualify the resultant entity as human-made. Fulfilling a criterial standard need not imply an absolute category or a division between mutually exclusive realms. In short, fulfillment of one condition of patentability need forebode no drastic consequences for basic philosophical categories of thought.

The ease of release from protests of this order does little to quiet fears that the Court has disrupted through legal intervention our basic understanding of what life is. Life is special, however difficult it may be to say in what the specialness may consist. The Court, in upholding the ruling of the lower court, has removed that specialness and voided the distinction between life and nonlife. As Ted Howard expressed it, "Life has no 'vital' or sacred property," according to the rulings and the briefs submitted in support of the winning position. Howard went on, "When those in power blithely call life 'a machine' or 'an industrial product' or a 'tool of manufacture,' then we had best take note of what is happening in our world." [32] These expressions of fear only extend the concerns registered for some time by the People's Business Commission, for whom Howard had spoken at the 1977 Academy Forum on DNA. On the prospects of extending DNA research into human genetic engineering, he remarked, "we have means to resist this final change of the human species. A storm of public outrage is coming, and it won't be gentlemanly." [33]

The fear that the sanctity and vitalness of life has been voided by the court and by DNA research emerges from biological ruminations over the difficulty in distinguishing the living from the nonliving at the level of microorganisms of certain orders. Statements to the effect that life is for the most part chemicals and that the gap between the living and inanimate has disappeared spawn anxiety in many people who are used to thinking in clear categorical terms. To erase a clear dividing line is for many to fuse categories and all that may apply ethically to those categories. Although it is fairly easy to demonstrate that nothing of the sort need necessarily follow, the convincingness of the demonstration for those absorbed in their fears is dubious.

Of course, in the West at least, life itself is far from universally sacred. We regularly and purposefully kill microorganisms; cleanliness and hygiene demand it. Medicine is partly built around killing germs that infect us. And our purposeful destruction of the living does not stop at the microbial level. Insect and rodent pests suffer under the swatters, spray cans, and poisons. Likewise, we do not treat with total impunity inanimate objects. Indeed, our growing respect for natural resource minerals seems almost proportional to our diminshing respect for the personal property of others. In short, the dividing line between the living and the nonliving says little or nothing to our ethical obligations with respect to the entities about us.

At the same time, we do recognize extensive ethical obligations toward many sorts of plants and animals (especially). Environmentalists and ecologists extend our obligations to all matter comprising the present and future ecosphere. We do not kill animals, domestic or wild, with total abandon, although we may debate about the permissible numbers, purposes, and conditions of killing. The humane treatment of animals is a well-engrained value within the culture, whatever questions may exist concerning specific actions we take in realizing the value. None of these values and obligations is affected by either the Supreme Court decision or reflections upon the difficulty of distinguishing a mere organic chemical from a microorganism.

Perhaps the key alteration in our ethical conceptions occurs when we move from the animal to the human plane. Arguments for animal rights aside, humans have rights, and obligations occasioned by rights, that go far beyond those obligations we have toward animals and plants. If sanctity of life has meaning anywhere, it is most meaningful when applied to humans. Kant used the idea of worth to distinguish human value from the means–end value relationship we apply to other things. Unlike price and other means of exchange, worth is not measurable or comparable. The axiologist, Robert Hartmann, spoke of intrinsic value in similar terms. And our religious traditions voice related ideas, whatever the leading concept, such as spirit or soul. Nonetheless, we do not yet always treat humans with the sanctity that supposedly inheres in their being human. Apart from obvious cases of cruelty and violation, we continually re-examine and find wanting our regular actions toward humans. There would be no cause for the existence of OSHA rules or regulations protecting the rights of the human subjects of experimentation were our observation of obligations occasioned by the rights of humans qua humans as thorough as our theoretical and philosophical protestations.

These remarks are but a set of reminders that with respect to the ethical dimensions of life, the living, and living, absolute categories distinguishing this or that contribute nothing at all. Although misinformed nostalgia for simpler times with less complex knowledge may provide an urge to stop research that further complicates our decisions and actions, no such possibility exists. Individual ethical decisions, while traceable in part to a variety of principles, never did nor will rest upon absolute distinctions and unalterable categories. We require instead sensitivity and carefully weighed rational and compassionate decisions that reflect as fully as possible our fundamental principles of

action, value, and responsibility. There is no release from the risk of
error in any decision we make; hence we shall likely never be free of
the need for mechanisms of justice and recompense. Various schema
derived from ethical theory, decision theory, and similar sources may
assist us in making better decisions, especially where complex factors
and large social consequences hang in the balance, but they remove
neither the risk nor the duty to proclaim when an action has gone too
far.

If our conception of life operates anywhere, it is within the main
line of common experience, not along the frontiers that separate one
concept from another. Thus, fear and not clear ethical reasoning in-
forms visions of domino effects toppling the sanctity of human life as a
result of declaring the patentability of a laboratory produced microorganism. Allowable extensions of the technology of recombinant DNA
research will require specific decisions as they become feasible, and
these decisions will be reached within existent but improvable social
institutions for such purposes. The recent Supreme Court decision did
not change our conception of the living, mostly because the
conception—especially in its ethical dimensions—never was clear
enough for a microbial decision to change. However, perhaps the
Court decision has provided the occasion for serious and useful reflection upon the concept of life, reflection that may assist us later to make
the important decisions concerning future uses of recombinant DNA
research when it has passed through the narrow entrance it has won
into microbiological knowledge.

Acknowledgment

I should like to acknowledge with gratitude the research assistance of
Ms. Kathleen Merek on this project.

Notes and References

[1] "The Right to Patent Life," *Newsweek* 45 (June 30, 1980), 74–75.
[2] United States Patent #4,237,224 (December 2, 1980).
[3] "Gene Goldrush Splits Harvard, Worries Brokers," *Science* 210 (November 21, 1980), 878–879; "Harvard Finally Backs Off Gene Venture,"
Nature 288 (November 27, 1980), 311.
[4] "California University Sues the Genetic Engineers," *New Scientist* 88
(November 27, 1980), 556.

[5]"Congress Shares Out Patent Licenses," *Nature* 288 (December 11, 1980), 527; "US Compromises on Genetic Engineering Controls," *New Scientist* 90 (April 30, 1981), 268.

[6]"Backlash Against DNA Ventures?" *Nature* 288 (November 20, 1980), 203–204.

[7]"The Right to Patent Life," *Newsweek, op. cit.,* 74–75.

[8]Diamond, Commissioner of Patents and Trademarks, v. Chakrabarty.

[9]See "Should Academics Make Money Outside?" *Nature* 286 (July 24, 1980), 319–320.

[10]Barbara J. Culliton, "Biomedical Research Enters the Marketplace," *New England Journal of Medicine* 304 (May 14, 1981), 1196–1197.

[11]"Harvard Backs Off Recombinant DNA," *Nature* 288 (December 4, 1980), 423.

[12]Bernard D. Davis, "Profit Sharing Between Professors and the University?" *New England Journal of Medicine* 304 (May 14, 1981), 1234.

[13]"Congress Shares Out Patent Licenses," *Nature, op. cit.,* 527.

[14]Vincent W. Franco, "Ethical Analysis of the Risk-Benefit Issue in Recombinant DNA Research and Technology," *Ethics in Science and Medicine* 7 (1980), 147–158.

[15]"US Compromises on Genetic Engineering Controls," *New Scientist, op. cit.,* 268.

[16]US Patent #4,237,224.

[17]*Federal Register,* 46 (October 30, 1981), 53984–53985.

[18]Daniel Callahan, "Recombinant DNA: Science and the Public," *Hastings Center Report* 7 (April, 1977), 20–23.

[19]Tabitha M. Powledge, "Recombinant DNA: The Argument Shifts," *Hastings Center Report* 7 (April, 1977), 18.

[20]Key Dismukes, "Recombinant DNA: A Proposal for Regulation," *Hastings Center Report* 7 (April, 1977), 29–30.

[21]"Backlash Against DNA Ventures?" *Nature, op. cit.,* 203.

[22]Culliton, "Biomedical Research Enters the Marketplace," *op. cit.,* 1201.

[23]*Ibid.,* 1198.

[24]Should Academics Make Money Outside?" *Nature, op. cit.,* 320.

[25]"Harvard Backs Off Recombinant DNA," *Nature, op. cit.,* 423–424.

[26]David H. Smith, "Scientific Knowledge and Forbidden Truths," *Hastings Center Report* 8 (December, 1978), 30–34.

[27]See, e.g., "The Right to Patent Life," *Newsweek, op. cit.,* 75; or Culliton, "Biomedical Research Enters the Marketplace," *op. cit.,* 1199–1200.

[28]Key Dismukes, "Life is Patently Not Human-Made," *Hastings Center Report* 10 (October, 1980), 12.

[29]Diamond v. Chakrabarty.

[30]Lee Ehrman and Joe Grossfield, "What is Natural, What is Not?" *Hastings Center Report* 10 (October, 1980), 10–11.

[31]Key Dismukes, "Life is Patently Not Human-Made," *Hastings Center Report* 10 (October, 1980), 11–12.

[32]Ted Howard, "Patenting Life," *The Progressive* 43 (September, 1979), 37.

[33]Quoted in Powledge, "Recombinant DNA: The Argument Shifts," *op. cit.*, 18.

Ethical Issues Raised by the Patenting of New Forms of Life

James Muyskens

The June 1980 decision of the US Supreme Court that a living organism could be patented under the provisions of the US Patent Act vividly symbolizes our entering a new era in the relationship between basic biological research and commercial ventures. As such it provides an excellent occasion to pause and reflect upon where we are going and whether we want to go there.

With the current revival of the spirit of Coolidge ("the business of America is business") it may seem close to heresy to ask whether there should be *commercial* exploitation of new forms of life. Yet the question must be confronted. Does the special character of the research, its funding source, as well as the products of the research call for other means of development—e.g., a nonprofit consumer, or citizen, or government-controlled or -regulated organization? Can a case be made that the gift of life should be reserved for the common good and not be allowed to be used for merely commercial or private gain? Willingness to consider such an alternative, of course, requires the willingness to engage in a thorough re-examination of our current economic system, its institutions, and its assumptions.

Before we manipulate and use life forms with the same abandon as we do inanimate matter we should consider whether this violates something important expressed by the age-old distinction between animate and inanimate matter. Perhaps our natural awe and respect for the special character of life should constrain us from allowing new forms of life, too, to become subject to restrictions of ownership by individuals, which is likely to lead to legal protection of *exclusive* control of these life forms by the fortunate few. Instead, if we reflect upon it, we

are likely to agree that life—given the high regard held for it through
the ages—should be set apart for special use to enhance the quality of
life for all and to diminish suffering and pain. Does not life hold a special place in the hierarchy of nature? Surely we are not prepared to say
that the distinction between the inanimate and the animate is irrelevant
when it comes to the question of our proper relationship to nature.
Somewhere, then, we must draw the line. We cannot agree to have
these new forms of life put to *any* commercially profitable or militarily
advantageous use to which inanimate matter is put. The use of new
forms of life to wage warfare, for example, would be particularly ob-
scene. Even for those who have lost a religious perspective, does not
life itself still seem to be sacred in the sense that it especially should be
immune from destructive and exploitative use?

At the outset we must be aware that granting exclusive control
over these new primitive forms of life may be setting an unfortunate
precedent for the way in which we allow control and use of more and
more complex forms of life. If we find repugnant the prospect of letting
anyone who is ingenious or wealthy enough have exclusive rights and
control over any forms of life they can obtain, where can we non-
arbitrarily draw the line? Should the line for such use be drawn at life
itself?

These remain interesting and difficult questions that must be
addressed—despite the likelihood that they can never really be answered. Consideration of the moral issues raised by patenting new life
forms, however, must move from this abstract and general level and
take as a point of departure the fact that we have embarked on the commercial exploitation and patenting of new forms of life. Whatever our
answer to the broader questions of the appropriateness of what we are
doing, the urgent questions are: (1) Who should enjoy the lion's share
of the profits and the major benefits that are likely to ensue from the
commercial development of new forms of life? (2) How can we minimize the unavoidable losses that flow from the marriage (already con-
summated) between university-based biological research and new
commercial ventures?

The importance of the Supreme Court decision concerning the
patenting of new forms of life lies more in the questions it raises than in
any practical clarity or direction it provides commerce. The stated purpose for granting patents is to promote the progress of science and
technology. The mechanism for doing this is to secure for a limited
time to inventors exclusive rights to their inventions and thereby protect them from competition that could deny them the chance of en-
joying the fruits of their labor. Whether granting patents will really of-

fer protection and accelerate commercial development of, for example, strains of bacteria that can make insulin and hormones, microorganisms that can mine silver and gold, and plants that can manufacture their own fertilizer, is by no means certain. The patent protection may be unnecessary or, on the other hand, may be no real protection at all. The commercial development of these products may proceed just as rapidly or just as slowly without patent protection. These are empirical issues concerning which there is no general agreement. The court action does not contribute much to their resolution.

What the court decision does, however, is to raise the fundamental question of the appropriateness of the various assumptions that are being made and steps that are being taken as we gear up for the full commercial exploitation of new forms of life. May we control life for personal gain? Should we grant someone exclusive ownership of some part of the animal or vegetable kingdom for a limited number of years? Is it right that a few individuals and companies reap the benefits of years of research sponsored by all of society through its tax dollars? To these questions we now turn.

Have we made an incorrect assumption if we consider a form of life (no matter how much human manipulation was required to fabricate it) to be a *human*-made invention? Thinking of these life forms as human inventions is, of course, a presupposition of patenting them. But can we with accuracy say that even the most intricate manipulation required to produce, say, a new bacterium with markedly different characteristics from any found in nature is *creating* it? As Key Dismukes has argued, to call such intervention in the normal processes by which strains of bacteria exchange genetic information human handiwork "wildly exaggerates human power and displays the same hubris and ignorance of biology that have had such devastating impact on the ecology of our planet."[1] The history of medicine and technology is replete with such arrogance. This arrogance leads to the notion that one's understanding of the fundamental mechanisms of life is complete and that one is in a position to create nature in one's own image and after one's own desires. We yearn for such control of life. Exaggeration of the human contribution in the development of new forms of life nourishes this hunger.

Exaggeration of a particular individual's contribution to the development of a new form of life puts that person in a stronger position to demand exclusive rights to "his or her creation." It is readily evident, then, why this, too, is a common tendency. What more effective rationalization for claiming exclusive rights to x than to convince oneself and others that x is the result of one's *own* effort. Many wealthy and successful people have had no trouble convincing themselves that their wealth and success are products of their own efforts, with little, if any, credit going to their parents who may have provided a sizable inheritance, their teachers who instilled discipline and guided them, or their society that provided the institutions, laws, and security essential for engaging in the activities by which they prospered. The self-made millionaire and the self-sufficient researcher are part of an American myth that should not be permitted to guide policy choices.

We must ask whether it is in our best collective interest to allow a few favored entrepreneurs and researchers to call certain segments of nature their own. Or when it comes to the development of new forms of life—a development, it is worth stressing, made possible by the contribution of all taxpayers—ought we to insist that their debt to society be recognized. Should we not as moral agents make certain that granting the right to anyone to exclude all others from making, using, or selling any organism (that is, granting a patent) is done only if it is clearly the best way to benefit society including the "lowest representative persons" of our society?

Patenting a new life form is different from ownership in ordinary property. Accepting the institution of private property does not entail acceptance of a right to own such things as new forms of life. Ownership in ordinary property is confined to specific, concrete, and tangible things. But the property rights sought in patenting of new forms of life are much more extensive than this. They encompass all individual members of the variety of the patented organism that exist during the term of the patent.[2] Surely society, acting through its legal arm, which grants patents, would be acting foolishly if it relinquished such broad rights to something in which it has heavily invested without a compelling state or community interest in doing so. The presumption must be in favor of these life forms remaining in the public domain (i.e., the burden of proof must be on those who wish to enforce restrictions of access).

The *traditional* ethos of medicine (one that medicine as big business has long since shed) according to which research is to serve humankind, rather than to constitute a vehicle for personal or corporate profit, is a good model to emulate. The skill of the physician was seen as a gift that was misused if not used altruistically and for the good of all (including the least advantaged). Especially since the development of new forms of life has arisen primarily from publicly supported, university-based research the debt to mankind must not be overlooked. Surely the fact that the basic research that makes these technological developments possible has had the federal government as its major sponsor adds weight to the obligation to use the results for the general

welfare. All the people through their tax dollars have paid for this research. On what basis then has any single individual or company the
right to cash in on the knowledge and technology generated with public
money?

Furthermore, as is the case with medicine and medical research, a
large proportion of the benefits flowing from the new forms of life are
in the areas of *basic human needs* such as health care and diet (food
production) and energy production. Medicine was considered an art set
apart from mere mundane activities because it could save lives and alleviate suffering. This skill was not to be withheld for reasons of politics, corporate profit or personal gain. The applications to which the
new forms of life can be put also save lives and alleviate suffering
[e.g., vaccine production (construction of specific bacterial strains able
to produce desired antigenic products so that killed or attenuated
disease-causing viruses need not be used), antibodies and hormones
(human interferon and insulin), specific cancer antigens, vaccine to
combat foot-and-mouth disease (a global plague that forces the slaugh-
ter of millions of animals each year and contributes to human malnutrition), genes for nitrogen fixation (a way to improve the crop yield per
acre that would do much to counter-act malnutrition), production of
pollution-free energy (e.g., algae producing hydrogen from water by
using sunlight) that would do much to lower the incidences of
pollution-caused disease]. Following the traditional medical model,
we should also see the work in genetic or biological engineering as set
apart and thus above the parochial constraints of politics, corporate
profit, and greed.

Justice (fairness) requires that this new knowledge and the resultant technology be used in ways that improve not only the general welfare, but especially the lives of the least fortunate in the society. As
John Rawls has argued in *A Theory of Justice,* social and economic
inequalities—if they are to be morally defensible—are to be arranged
so that they benefit the least advantaged.[3] Only if granting special
rights and benefits (e.g., the exclusive rights of ownership of a patent
and its monetary returns) to someone or some small group is the best
way of achieving fair distribution of a research product produced col-
lectively can such uneven distribution be defended. If any other arrangements resulting in less disparate distribution of benefits (yielding
greater gain to the lowest representative persons) would be as or more
efficient, the alternative especially profitable to the "inventors" and
stockholders is not morally defensible.

It is often uncritically assumed that the best way to achieve the
widest distribution of the products of research is by commercial exploitation. The mechanism of the free market, it is believed, will work best

to reach even the poorest among us. Of course, this faith is much more readily held by the affluent than by the needy. If your health and welfare depend on it, you readily recognize that many things can happen to prevent the "trickle down effect" from working. What assurance is there that it will be profitable for the commercial holders of patents to produce the products needed by those with very limited resources? The big profits may arise in producing only what the affluent can afford. Recent and expected reductions in the federal budget for such things as health care and social welfare, along with the adoption of an economic policy permitting high rates of unemployment, diminish further the prospects of profitability being an adequate base for delivering the benefits of the research to the least advantaged. Unless assured of being paid well by the government to work on health problems mainly affecting the poor (e.g., to produce a vaccine of special benefit to certain disadvantaged groups), a company wishing to maximize its profits is well-advised to turn to other projects and let the poor fend for themselves. (Ironically, if we had adopted a national health insurance program or a national health service, this objection could be met. The current move away from such ideas and the reaffirmation that health care delivery is to operate as much as possible on "free market" principles weakens the case for the development of these new health- and welfare-related technologies in accordance with relatively unrestrained market conditions.)

Since we cannot always expect profitability to coincide with social good and justice, the search for means of spreading the benefits of the research that do not rely primarily upon the profit motive is morally required. At the same time, in a profit-oriented society such as ours, there are few other ways to provide incentive. So it would be naive and foolish to remove altogether the incentive of profit. The task before us is trying to find the right mechanism for achieving the proper balance. Unless carefully circumscribed, the patenting of new forms of life is not the best mechanism for achieving this goal. Carefully circumscribed, it may be.

Let us return then to the question of the appropriateness of the granting of patents for these new forms of life. In addition to the reasons mentioned earlier why those involved in biological research and engineering find it tempting to consider the new forms of life they are producing as inventions and as their own creations is the conviction that an inventor has the "moral right" to reap the fruits of his or her labor and perhaps even to control, at least for a limited period, its production and use. This intuition conforms to the longstanding philosophical doctrine that affirms that a person has a natural right to the fruit of his or her labor. On John Locke's view of private property, for example, a person acquires a property right to things originally in the public domain when they are altered by that person's labor and efforts. The granting of patents is one way of providing the protection and force of law to this (putative) natural right.

I have no quarrel with this traditional view. The difficulty arises in its application to patenting of new life forms. We have already challenged the claim that these new life forms are inventions. If they are not, they do not satisfy one of the necessary conditions for patenting. However, even if they are, the idea that someone *by himself or herself* has invented a new form of life is simplistic. This notion of rugged individualism in scientific discovery is merely nostalgia for an earlier, simpler era. With rare exceptions today, the chance for profit from a promising development is a result of the support of public money, if for nothing more than the use of university facilities, laboratories, or computers. In one way or another, a large share of the venture capital has come from society itself (as, for example, through government funded grants). The "invention" is, in fact, the result of a partnership. When the partnership is successful, one of the partners is not entitled to the profits at the expense, or to the detriment, of the others. Those who have put up the capital quite rightly expect a reasonable return on their investment.

Offering a patent to someone when doing so denies a fair share of the profits to all those responsible for the invention cannot be defended. It is both unfair (as we have seen) and counterproductive (as we shall see). It will discourage the investment in basic research. The news items concerning university professors seizing the opportunity to convert their basic knowledge into consumer products and cold cash have made the questions of who should profit from government-funded research a sensitive political issue. What good reasons are there for the public to provide resources for research that reaps high profits upon a few investors, academics, and former academics while there are no tangible benefits for them now or perhaps even in the distant future? In these circumstances it is not surprising that an administration looking for places to slash the federal budget finds little opposition among the general public for sharp reductions in the budget for basic research. The golden opportunity to demonstrate the value of basic research has been lost amidst stories of sudden wealth and good investments for those with capital to invest.

Current trends are that money for basic research is growing scarcer and scarcer. The beleaguered National Science Foundation and National Institutes of Health have funded broad areas of basic research.

Although these organizations now have fewer tax dollars to work with, the Pentagon, which funds projects with identifiable value to the military (but not basic research as such), continues to spread its largess. Despite the fact that it is *most* doubtful that spending these vast sums for national defense really is in the common good, it is an argument that carries the day. Among the many reasons for this is the fact that with national defense expenditures (in contrast to so many other government programs), rich and poor alike are provided with an equal share in the product (protection). They are not given varying protection depending on their ability to pay or their access to venture capital. Had we been truly concerned with the just distribution of the products resulting from our expenditures on basic biological research in recent years, the society-at-large could recognize as well its contribution to the common good and its benefit for everyone. A commitment to the just distribution of these benefits is a first step toward popular support for restoring unrestricted federal funding for basic research in these areas.

Accepting the massive loss of relatively unrestricted federal dollars for research as a *fait accompli,* more and more universities have jumped into the arms of business. In order to maintain a high level of research activity, sources from business have been sought to replace that lost from government. Of course, businesses (quite rightly) are in business to make a profit. They (as well as the Pentagon) are not as interested in general, basic research as in very specific projects having at least the potential for practical application. The criterion for business interest is the potential for profit. It may seem that the solution to this problem is to bring the techniques of the marketplace into the laboratory. The current shortage of funds and the lack of general support for basic research may be seen as the occasion to turn adversity to the researchers' advantage. If the only choices are no research or research at the behest of industry with the researchers becoming industry's well-paid servants, we would choose the latter. Some research going on is probably better even for the "lowest representative persons" than having none at all.

Nevertheless, this solution—and the developments in its direction—open up a Pandora's box for the staid world of the academic who traditionally has scorned using the knowledge gained from research to make money (beyond that of cadging grants for further research, royalties, or consulting fees). Reputable universities have carefully limited and monitored any outside activities professors are pursuing. An unhappy exception has been the practice of some medical schools, which permit lucrative arrangements whereby a member of the medical faculty can draw full salary while also maintaining a full-time practice. If this is the precedent for the biologists now heading their own companies while retaining university standing, universities will certainly be the losers.

The most obvious problem is one of getting full-time service from someone with outside commitments. Yet in many cases it may be that the prestige of the faculty member and thus the benefits of the affiliation will be great enough to compensate for the little time he or she can give to the university. A more serious problem is the very real and already evident potential for conflict of interest on the part of faculty members with external commercial interests. According to the *New York Times,* for example, at Harvard and Stanford "there are already reports of intense rivalries among laboratories, led by professors with different business interests, that are working on competing projects. Graduate students in some cases are said to be sworn to secrecy about the nature of their work" (February 21, 1982).

Clearly there is the potential for conflict of interest in the business–university connection. For example, as the *Times* report indicates, would not an individual scientist with a corporate connection be inclined to withhold information he or she has developed in order to protect the profits from the company's commercialization of those findings? Someone on the board of the company may even feel the scientist, as employee, has a *duty* to keep his or her findings secret to protect the company's interests. However, such secrecy runs contrary to a long-standing and honorable tenet of the ethics of research within the university, namely, the publication and dissemination of research results for free use by one's colleagues. In accepting the role of research scientist, one accepts as one's own this obligation to one's research colleagues. We see then that the duties that arise from these dual roles are not compatible. A scientist cannot serve these two masters.

If there is to be a close relationship between the university and business, some more complex arrangement that resolves the problem of dual loyalty must be devised. This can be done in a variety of ways, some more likely to be successful than others. Any approach, however, that fails to keep university researchers and administrators out of the business of making a profit is incompatible with the role of the university and that of the scientist within it. On the other hand, if the university does not protect its interests—including that of having funds for further basic research—it will not be carrying out its mandate to pursue knowledge not only for its own sake but also for the benefit of mankind.

If the work being done in university laboratories directly or indirectly yields high profits, one of the best ways to insure that it be used to benefit the public at large is to have it put back into the pool for continued research. Patenting may be one of the means necessary to accomplish this. The proper functioning of a board or faculty committee that would maintain its commitment to the traditional values of the free flow of information and the pursuit of knowledge for its own sake is also required.

A university setting up its own company with outside venture capital and with the university given a share of the stock (such as the Harvard University proposal rejected by its faculty last year) does not meet these requirements. Such an arrangement does not maintain the clarity of distinction between university research and commercial development that avoids conflicts of interest. Related proposals now under review at several universities that do more to separate research faculty and university administrators from business pressures and the pursuit of profit may satisfy these conditions.

With regard to industry-subsidized research within a university, some compromise on the principles of secrecy and the free flow of information may be acceptable. For example, at Yale University, one of the compromises found acceptable by their faculty committee on cooperative research, patents, and licensing was that Yale researchers agreed to keep their findings secret for up to 45 days, if requested to do so by the company. However, no restraints were allowed on publication thereafter (*New York Times,* February 18, 1982). If industry generally is willing to accept such a compromise, and if these industry-subsidized projects are but a small part of a university's full research program (and all such agreements are made in the open), the risk to academic freedom and the free-flow of information by industry-subsidized research is minimized.

The need for the free-flow of information in university research as it is traditionally done cannot be denied. In the rush to obtain money from industry and to secure patents to protect the results of one's research, there is a danger of distorting the scientific process. "By its nature science must be conducted openly in order to work well. Peer review, wide dissemination of findings, and open debate are essential mechanisms for obtaining objectivity and consensus."[4]

In addition to concern about the quality of research once it is too closely mingled with profit, the issue of the public's perception of the scientifc community's commitment to objectivity must be considered. Someone standing to benefit financially from some line of research is not likely to be seen as the right person to ask about its safety, risks, potential benefits, or whether it should be preferred over some other line of research. The disinterested expert may become an endangered species unless we guard against these inroads. Without careful controls, universities themselves could appear to be and perhaps even become mere servants of capitalism. The greatness of the university has been that it stands above these struggles.

Some may wish to argue that these constraints on the university–industry connection are all premised on a somewhat naive view of today's universities. Universities have entered into all manner of alliances with grant-giving agencies and have accepted numerous conditions for receiving grants. For years universities have been willing to compromise and accept conditions limiting the free-flow of information and raising the spectre of conflict of interest—all in return for much needed money. Why should we treat cases of research on new forms of life any differently?

The problem is that we have moved gradually and often uncritically in this direction. The decisions that have been made were often made on the basis of expediency and were not intended as precedents. With the vast new opportunities for research having commercial application coupled with the cutback in federal funds with relatively few strings attached, the potential for serious damage to university research has greatly increased. We stand in danger of a wholesale change in the way basic research will be funded and conducted. This change, even if beneficial to most parties in the short run, will be detrimental to university research and also industry in the long run. It will undermine loyalty to colleagues, a sense of community, the commitment to objectivity, and the pursuit of knowledge for its own sake.

The current rush to patent new life forms ["The United States Patent Office has a backlog of more than 100 applications for patents on living organisms and on procedures related to genetic engineering" (*New York Times,* June 22, 1980)] is the point to stop and turn again to the traditional ideals. More than ever, it is essential that we draw the distinction very clearly between the purpose of research in the university and the purposes of research in industry. Those doing research in the university must renew their commitment to the traditional ethics of research and actively work to promote means to get the benefits of their work to the society-at-large, including the least advantaged. Quite simply, when the patenting of new forms of life (by universities, individuals, or industry) will secure (directly and indirectly) the benefits for humanity, patenting should be encouraged. When patenting will likely benefit the favored few with little regard for the rest, requests for patents should be denied.

The main moral question raised by the patenting of new forms of life is that of the constraints and limits we should place on the commercial exploitation of them. Factors supporting constraints are: the special status of life, the public funding of the research producing the new life forms, the requirements of justice in the distribution of primary goods, and the role of the university in the development of the new forms of life. Any commercial development and patenting of them must not violate any of the following conditions: (1) respect for life as something to be reserved for the common good, (2) fairness to all the taxpayers who have underwritten the research, (3) distributive justice concerning goods essential to life and well-being, which includes the requirement that, if anyone gains, the lowest representative person must gain from the distribution, and (4) professional research ethics that commits university-based scientists to engage in the free exchange of scientific information and the pursuit of knowledge for its own sake and for the betterment of mankind.

Notes and References

[1] Key Dismukes, "Life is Patently Not Human-Made," *Hastings Center Report* 10 (October 1980), p. 12.

[2] Irving Holtzman, "Patenting Certain Forms of Life: A Moral Justification," *Hastings Center Report* 9 (June 1979), p. 11.

[3] John Rawls, *A Theory of Justice* (Cambridge, MA.: The Belknap Press of Harvard University Press, 1971), p. 83.

[4] Dismukes, *op. cit.*, p. 12.

Index

Index

C

Callahan, Daniel, 173
Campbell, A. G. M., 37
Case study of health care allocation, 108ff.
Categorization of behavior, 133
Cebik, L. B., 163
Chakrabarty process, 165
Christianity and killing, 31ff., 41
Cohen-Boyer patent, 165, 171
Community Health Care Plan, 111
Competence, 138, 143
Conflict of interest laws, 169
Consent, 138
    informed consent doctrine, 142–143
Cost-cutting practices, 123
Culliton, Barbara, 169

D

Dangerousness, 140
    prediction of, 139
    procedural issues regarding, 141
    prospective, 140
    retrospective, 140
    substantive issues regarding, 141
Davis, Bernard, 170
Defective infants, 37ff.
Deinstitutionalization movement, 144
Demythologizing reproduction, 88
Deprivation of liberty, 139–140
Deterrence, 152
    limited deterrence theory, 154
Diamond v. Chakrabarty, 167, 179
Dismukes, Key, 173, 189
Distribution of health care, 91–106, 116
Donaldson v. O'Connor, 136, 142
Duff, Raymond, 37
Duty to be cured, 137
Duty to be treated, 137

U

W

Z